SHOP

设计速递 **专卖店设计**

精品文化工作室 / 编
赵欣 / 译

大连理工大学出版社
Dalian University of Technology Press

图书在版编目(CIP)数据

专卖店设计：汉英对照 / 精品文化工作室编；赵
欣译. — 大连：大连理工大学出版社，2012.5
（设计速递）
ISBN 978-7-5611-6816-5

Ⅰ. ①专… Ⅱ. ①精… ②赵… Ⅲ. ①专卖—商店—
室内装饰设计—作品集—世界 Ⅳ. ①TU247.2

中国版本图书馆CIP数据核字（2012）第049304号

出版发行：大连理工大学出版社
（地址：大连市软件园路80号　邮编：116023）
印　　刷：利丰雅高印刷（深圳）有限公司
幅面尺寸：225mm×300mm
印　　张：15
插　　页：4
出版时间：2012年5月第1版
印刷时间：2012年5月第1次印刷
责任编辑：刘　蓉
责任校对：李　雪
封面设计：四季设计

ISBN 978-7-5611-6816-5
定　　价：210.00元

电　话：0411-84708842
传　真：0411-84701466
邮　购：0411-84703636
E-mail：designbooks_dutp@yahoo.cn
URL：http://www.dutp.cn

如有质量问题请联系出版中心：（0411）84709246　84709043

SHOP

专卖店设计

Contents 目 录

==========CLOTHING 服装

==========ACCESSORY 配饰

LIVING 家居

OTHERS 其他

CLOTHING

服装

地点	面积	设计师	设计公司
/ 北 京	/ 2000m²	/ 迫庆一郎	/ SAKO建筑设计工社

Beijing Ripple

北京波纹

This is a clothing photography exhibition hall for seven brands. The seven brands introduce "event" as the theme, showing their products to the world in the 14m high and 60m wide space. All of the professional winter clothing brands choose "winter" and "snow" as the theme, introducing various ways of displaying in the space of more than 2000m², to coordinate well with the space.

Downstairs is the exhibition hall of various brands, and four sub-brands of "snow image" echo with the adjacent space through "door-style frame". The upstairs is a multi-functional space with center of performance and exhibition, and cafe and office areas, each area takes the maximization and optimization of utilization into account, so each space fully plays its role. The visitor can feel the amazing effect, and the brand effect wins support among people.

New Brand

Cafe

Snowimage Women

VLASTA

MISS SUN

Snowimage Men

Snowimage JUNIOR

4F plan　　0　　5　　10　　　　20m

Exhibition Space

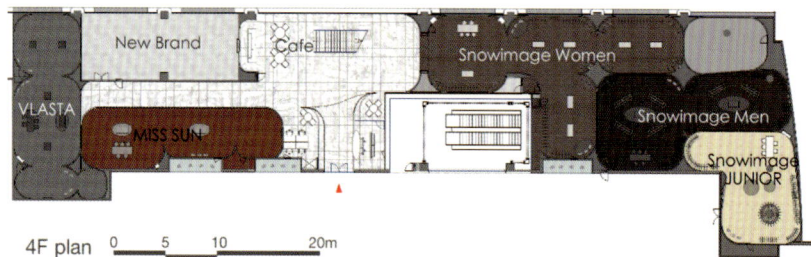

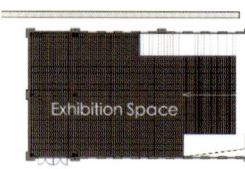

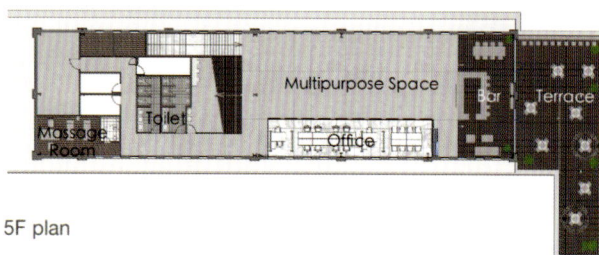

5F mezzanine plan

Multipurpose Space

Massage Room

Toilet

Office

Bar

Terrace

5F plan

这是七个品牌的服饰摄影展示场。七个品牌以"事件"为主题，在高14米、宽60米的室内向世界展出各品牌的商品。所有的专业冬装品牌，以"冬天"和"雪"为主题，在2000多平方米的空间里运用了多种展示方式，以便与空间很好地协调在一起。

楼下是各品牌的展示厅，"snow image"的四家附属品牌通过"门式框架"呼应相邻空间。楼上是以表演和展示为中心的多功能空间以及咖啡区和办公区，各区域都充分考虑到利用率的最优化和最大化问题，让每一个空间都能充分发挥其作用，让人产生惊艳效果的同时，也让品牌效应深入人心。

地 点	面 积	设计师	设计公司
/武 汉	/ 850m²	/ 迫庆一郎	/ SAKO建筑设计工社

Jeanswest in Wuhan

__ 武汉真维斯

This is a flagship store of Jeanwest in Wuhan. Jeanwest is worthy of the name "big shot" in Chinese casual clothes industry, and the store design is mainly simple and fashionable. The simple design is coupled with modern shaping materials to create a cool and comfortable shopping environment for modern urban youths,

The store is mixed with men's and women's clothes, and the store design is based on the brand's personality under free and casual route. The whole space features only one color, pure white, but it is different from the ordinary flat white. The glass inner walls with different sizes are introduced as collage materials, coupled with lighting to create a unique ceiling wall. The facade style is matched with panels of different sizes to create a wall model with high sense of level, coupled with grid displaying frames of different thickness, the whole space features only one color, but it is still full and vivid.

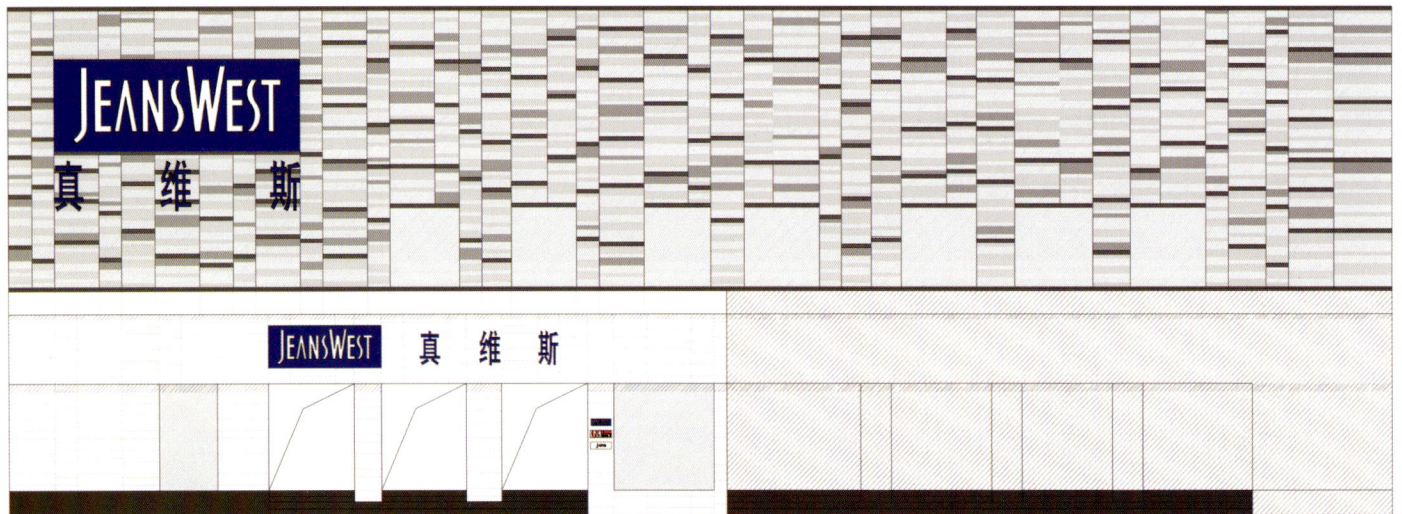

SOUTH ELEVATION

0.5   3   10(m)
0 1   5

TOP = 12,320mm

4100

3FL= 8,220mm

4000

12320

2FL = 4,220mm

4270

1FL= ±0mm

650

GL= -600mm

JEANSWEST

真 维 斯

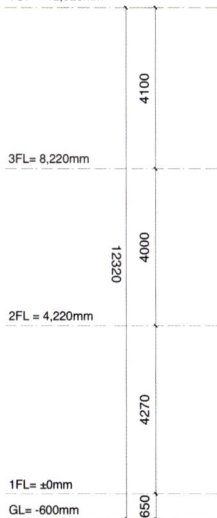

EAST ELEVATION

WEST ELEVATION

0.5   3   10(m)

0   1   5

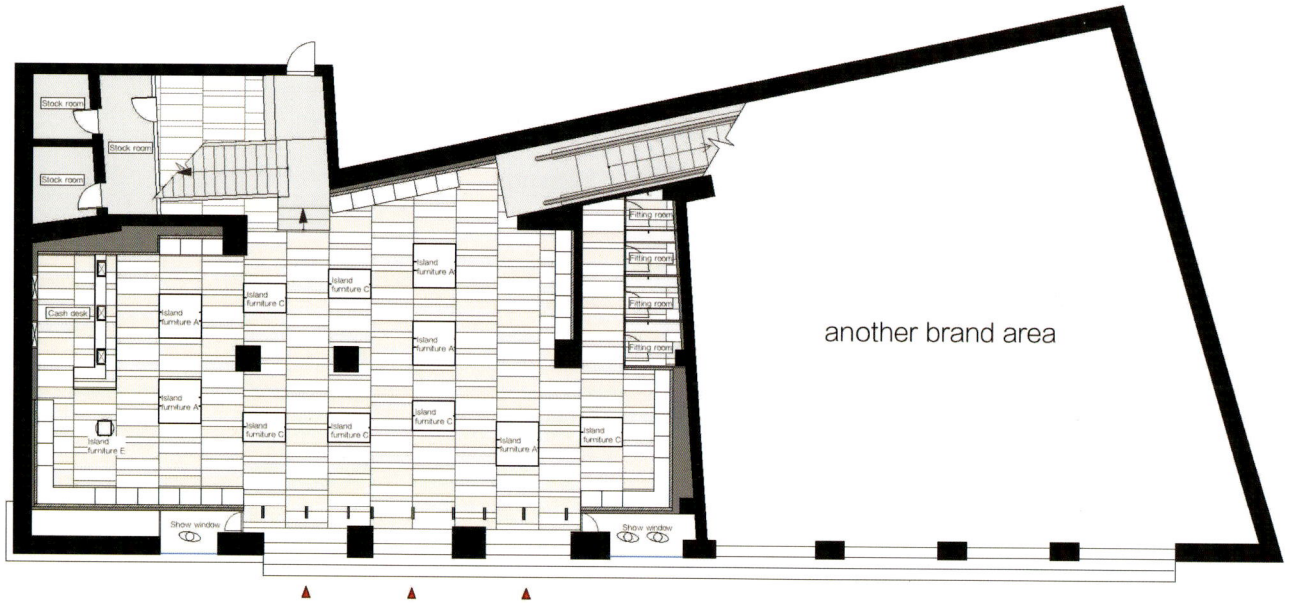

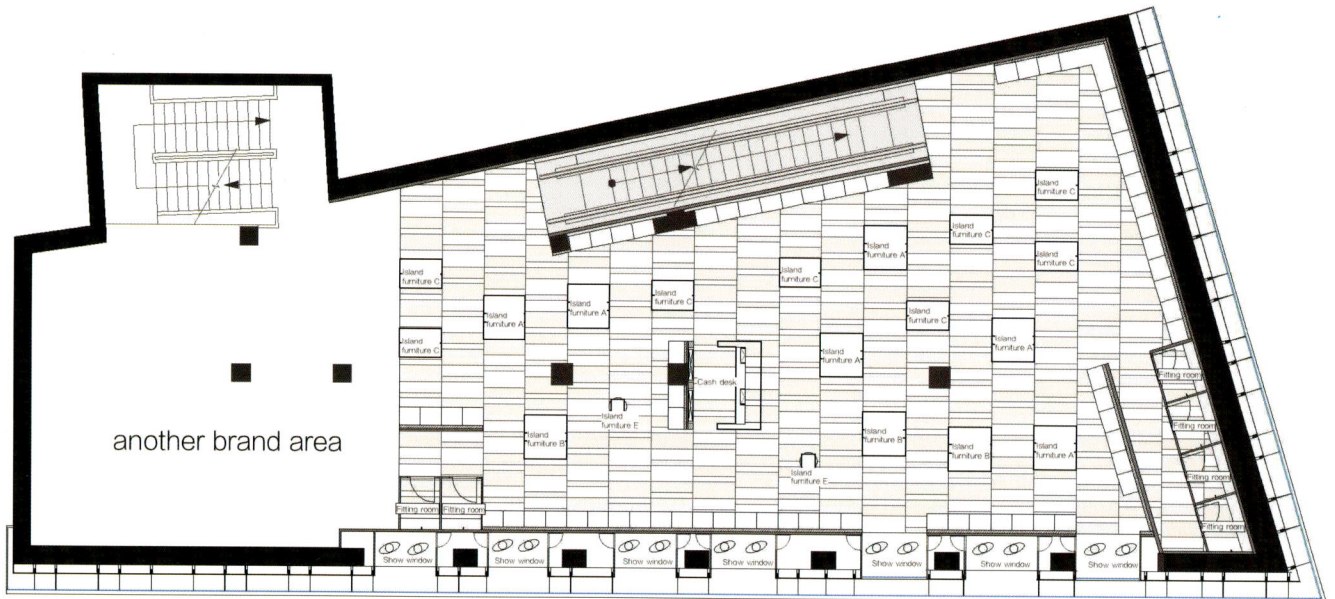

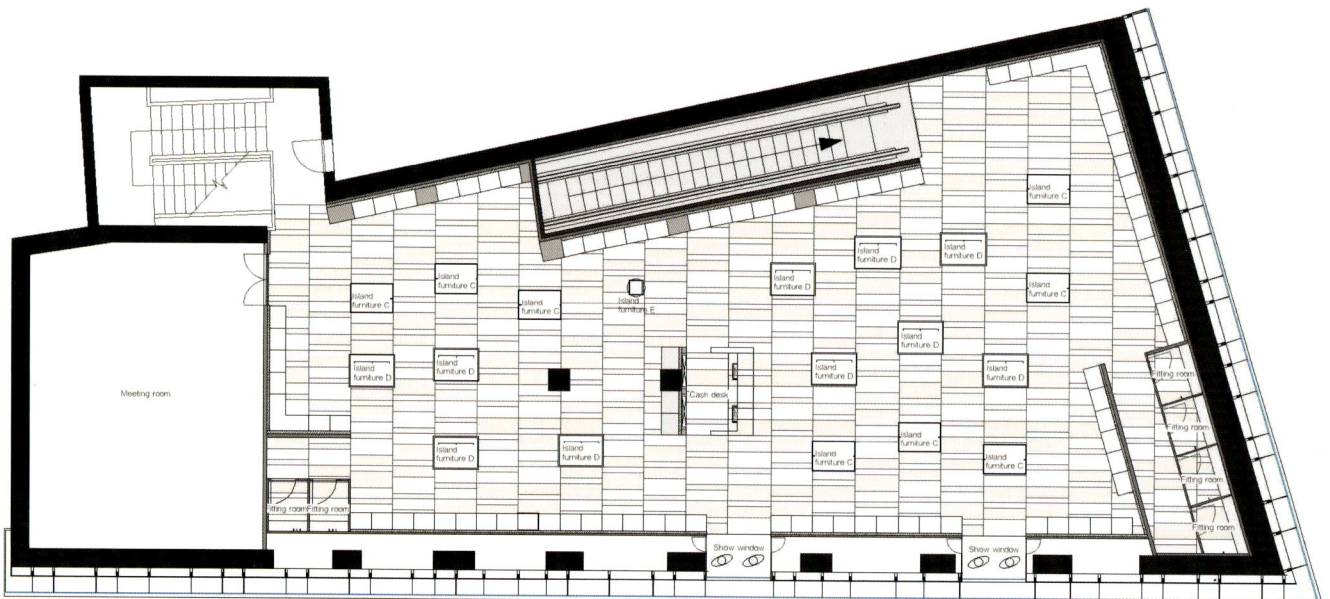

　　这是真维斯在武汉的旗舰店。真维斯作为中国休闲服装行业名副其实的"大鳄"，店面的设计以简约时尚为主调，以简约的设计配合现代型材，为现代都市年轻人创造了一个潇洒自在的购物环境。

　　本店是一个男女装混合卖场，店内的设计依据品牌的个性，走轻松自由的路线。整个空间只有一种色调：纯净的白。但是这种白却不同于普通的平板白，设计师用不同大小的玻璃内壁作为拼贴材料，配合灯饰打造出一面别具特色的天花墙。立面造型以大小不一的板面拼接打造出极具层次感的墙面造型，加上厚度不一的同色格架展示，整个空间虽然只有一种色调，却依然丰满、生动。

Puma Store

—— 彪马专卖店

The idea of the new store concept was to bring back the joy into retail environments while aiming at a sustainable and innovative retail design. Two floors and more than 200 square meters of shopping space are filled with a broad range of footwear, apparel and accessories including puma's black label featuring the brand's designer collaborations with hussein chalayan and alexander mcqueen. A six meters high brand wall in puma's iconic red is another eye catcher of the store.

The idea of a colorful and joyful retail experience is further transported into the changing rooms where a special "puma peepshow" – a red box that opens up exposing something unexpected being it video clips or product presentation – further allows customers to engage with the brand. Generally customer interaction is written in big letters at puma and therefore an important aspect of the store design including the integration of the latest communication channels: iPads are not only scattered around the shop attached to display tables but assembled at the "puma joy pad" – a huge iPad wall framed by red transparent glass – that allows customers to interact with puma using specially developed apps.

Despite numerous joyful elements the design concept sets a high value on sustainability. This is reflected in the use of ecological materials such as FSC certified wood and certified floor finishes, low-emitting paint and an efficient lighting concept mix to save energy consumption.

The store in Paris is a first step towards a generally novel retail experience. A rollout of the new concept in other countries is envisaged.

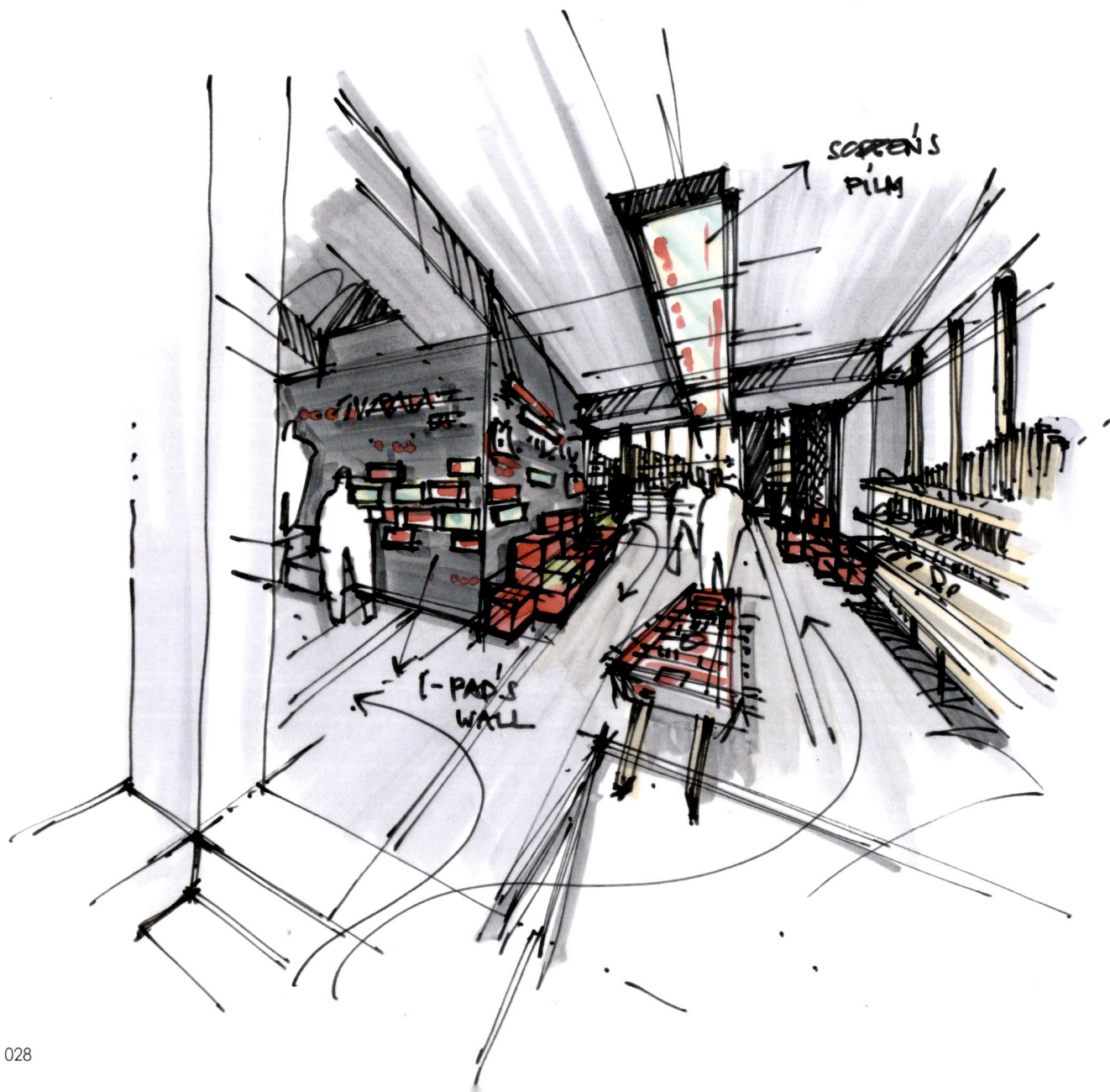

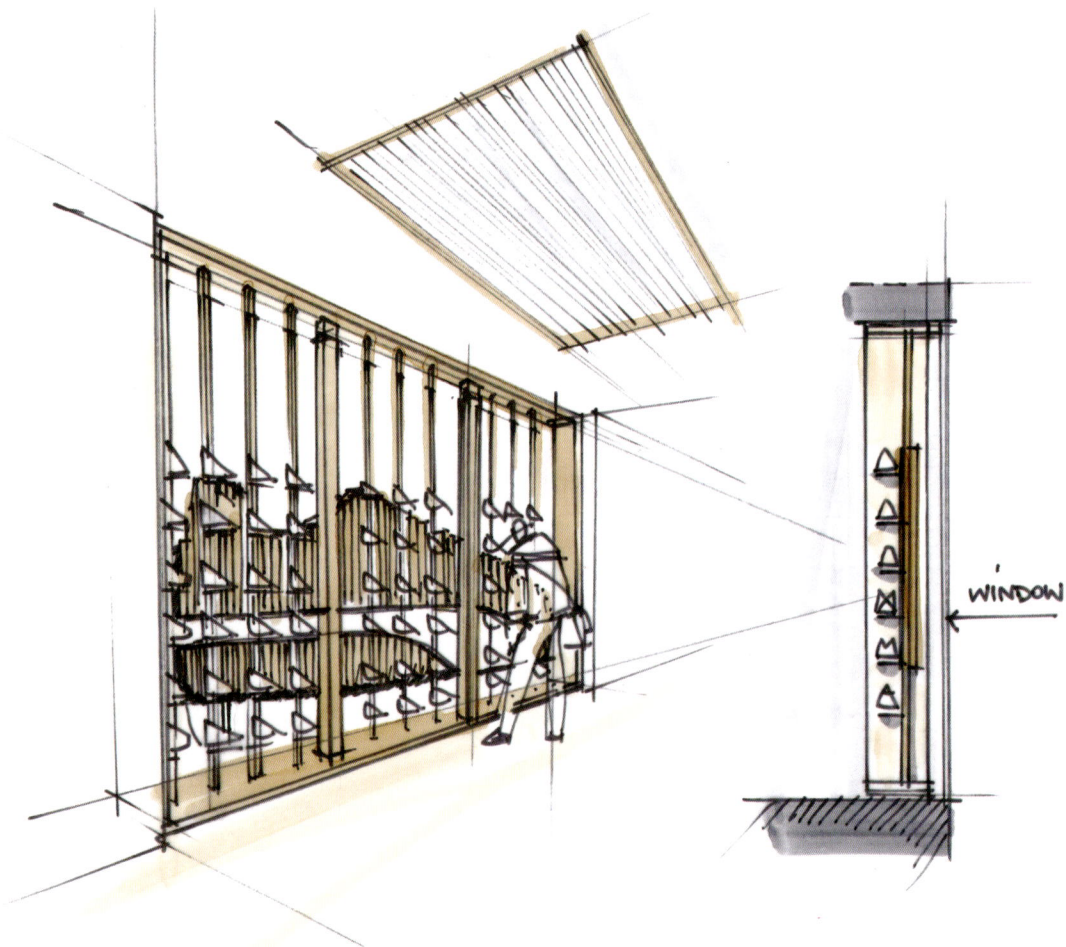

WINDOW

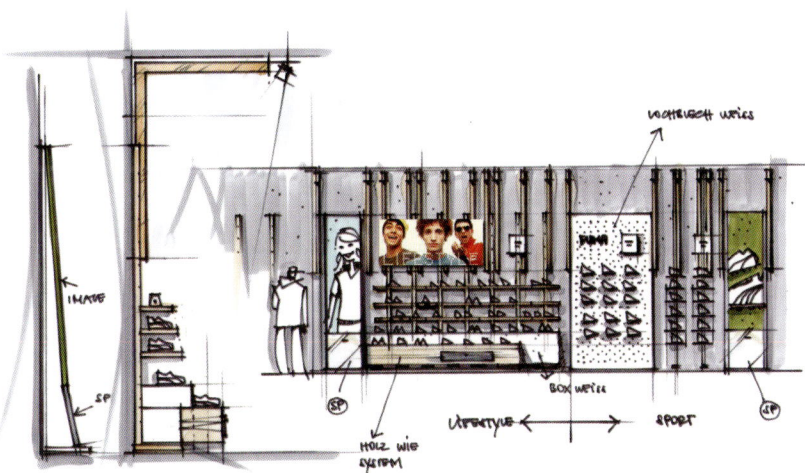

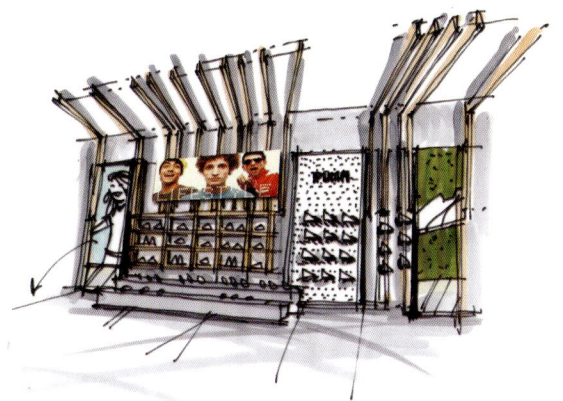

新概念店的想法是回归零售环境的喜悦，同时以保持可持续发展的创新型零售设计为目标。两层200多平方米的购物空间摆满了各式各样的鞋、服装及配饰，包括彪马的黑色商标，突出了品牌设计师与侯塞因·卡拉扬和亚历山大·麦昆的合作。六米高的品牌墙上设计了红色彪马标志，这也是吸引人们眼球的另一方面。

多彩欢乐零售体验的想法被进一步运用到试衣间中，特殊的"美洲豹偷窥"——一个向上开的红色盒子展示着令人意想不到的视频片段或产品介绍——使客户进一步融入到品牌中。一般来说，在彪马店里，客户互动要以大写字母显示，因此店面设计的一个重要方面是最新沟通渠道的融合：与显示台连接的iPad不仅散落在商店里，并且配有"彪马喜悦垫"——一个由红色透明玻璃构成的巨大iPad墙——客户可以利用专门开发的应用程序与彪马进行互动。

除了无数的欢乐元素外，设计理念还高度重视可持续发展，体现在环保材料的运用上，如FSC（森林管理委员会）认证的木材及地板饰面、低排放油漆和高效照明概念组合以节省能源消耗。

彪马巴黎专卖店迈出了一般新颖零售体验的第一步，新概念将会在其他国家一一亮相。

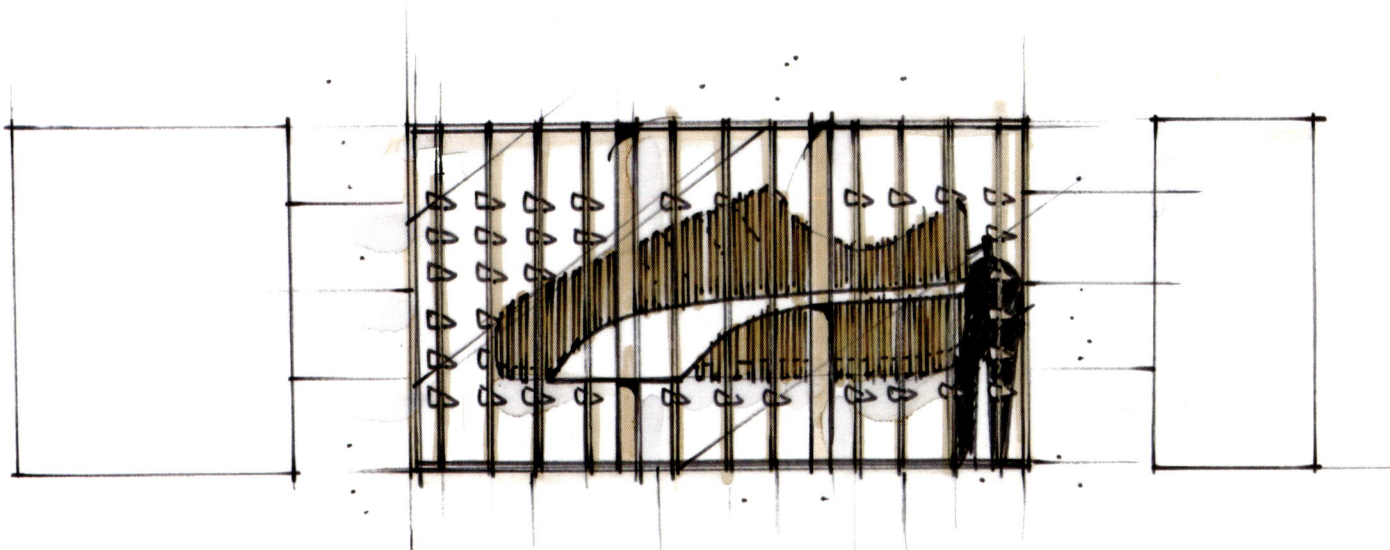

C'EST À LA
FIN DU BAL
QU'ON PAIE
LE MUSICIEN

À COUP
SÛR ELLE
TOMBE

À COUP
SÛR ELLE
TOMBE

地 点	面 积	设计公司	摄 影
/ 柏 林	/ 502m²	/ plajer & franz studio	/ diephotodesigner.de, ken schluchtmann

Galleries Lafayette

—— 拉斐特画廊

Galleries lafayette needed a new space for young girls fashion that not only transports the values of galleries lafayette but also gives 15 young jeans brands their own specified areas.

The building works with cones as overall concept. The cones penetrate the building to various depths. The one in the center goes all the way down to the basement. The character of galleries lafayette and the cone became the central point of concept. Circular movement and the game with light melts together the existing and the new without destroying unique characters for each space.

The new space is organized in different zones

1. The newly built circular stair to access the space. Round display discs are attached with bustforms. You will find them again together with the circular wall pattern on the escalator wall. This attracts attention on the lower floor leading you to the highlight area further up.

2. Along the walls of the new space a specially designed and highly flexible wall system offers an individual and inexpensive way to display the brands.

3. The main focus in the middle of the space is absolutely dominant and not branded. Here you can find a focus area with cross merchandising which changes to seasonal needs or special events. Nevertheless it harmonizes with the brand sections along the wall. Around the existing glass cone the concept implies the rotation of space with round discs. Supported by the white floor and ceiling discs this flexible space puts the fashion on the spot.

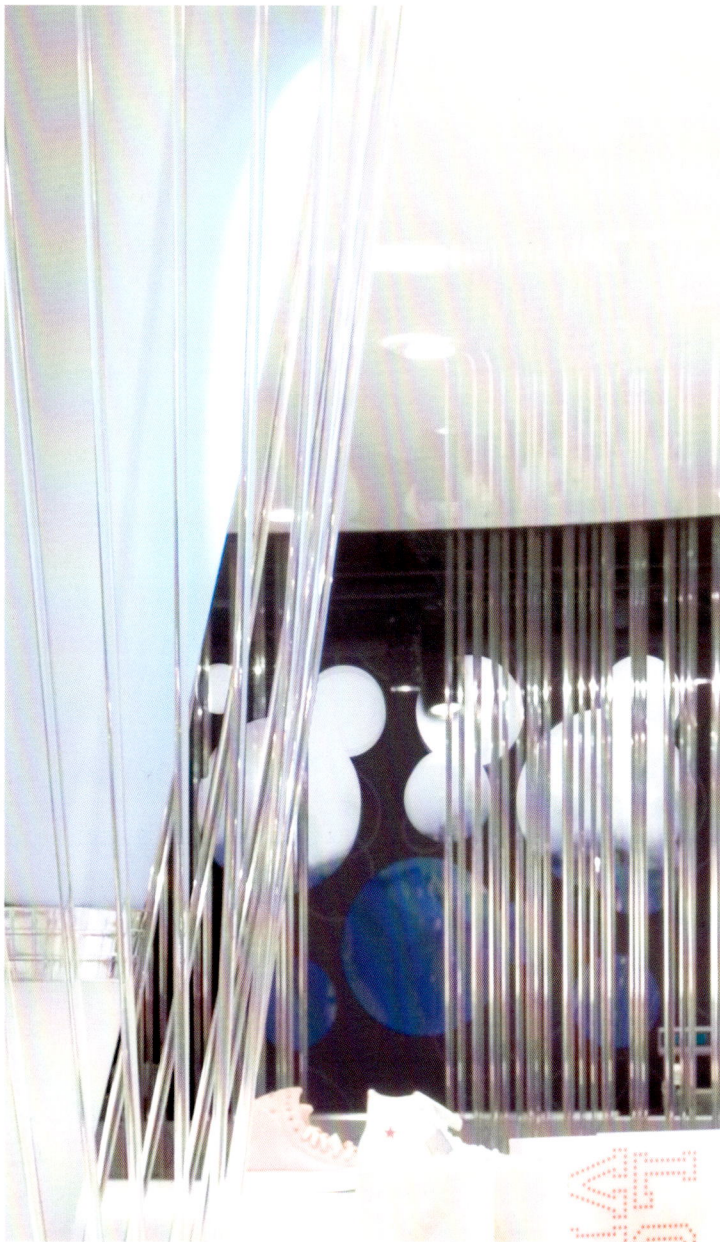

拉斐特画廊需要为时尚的年轻女孩创造一个新空间,它不仅要传承拉斐特画廊的价值,而且要为15个新兴的牛仔品牌提供专属区域。

该建筑以锥形作为整体概念,锥形贯穿整个建筑的不同层次。中心位置的一个锥形一直穿透至地下一层。拉斐特画廊的特色和锥形成为概念的中心点。循环运动和光效游戏融合在一起,形成现有的新区域,并没有破坏单个空间独有的特性。

新空间由不同区域组合而成。

1. 新建的圆形楼梯带领人们进入空间。圆形的展示光碟半掩着贴在墙上,视线向上,您还会在墙面上发现展示光碟与圆形墙壁的图案,此种景象吸引着人们从楼下走向楼上的精彩区域。

2. 沿着新空间的墙壁,一个特别设计的、高度灵活的墙壁系统提供了一个独立且廉价的品牌展示方式。

3. 中间空间的主要焦点是绝对优势,而非品牌化。这里,您会发现一个交叉销售的中心区域,是季节性需求或特殊事件区。即便如此,它仍与沿墙而设的品牌区和谐相处。围绕现有玻璃锥形,此概念暗示着空间顺着圆形光盘旋转。在白色地板及天花板光盘的映衬下,灵活的空间将时尚展现得淋漓尽致。

地点	面积	设计公司	摄 影
/ 汉诺威	/ 1543m²	/ plajer & franz studio	/ diephotodesigner.de, ken schluchtmann

s.Oliver /
New Store Concept

___ s.Oliver新概念店

As one of the most successful fashion and lifestyle brands in Europe, s.Oliver caters for all age groups and covers all fashion styles with its broad range of segments including "Casual", "QS" and "Selection". The segment s.Oliver Casual is characterized by natural and stimulating tactile surfaces that are combined with a warm color palette to create a cozy atmosphere.

Steel, dark colors and an open ceiling define "QS" for the young urban target group. Specifically designed wallpapers with graphic elements and altered wood finish designs show a clear segment that also appeals to the older crowd.

"s.Oliver Selection" constitutes the high-end and elegant business outfit. Metallic finishes are combined with black and white surfaces. The specially designed metallic wallpaper underscores that contrast between precious matte finishes and the luxury of high gloss elements such as the highlight wall made of mirror and black glass.

"s.Oliver Junior" caters to age groups between 0-14. Design elements from the casual area are used and redefined with new surface finishes. Brick wall cladding, fresh playful elements such as flowers for babies and bright colored furniture are characteristic of the youngest segment.

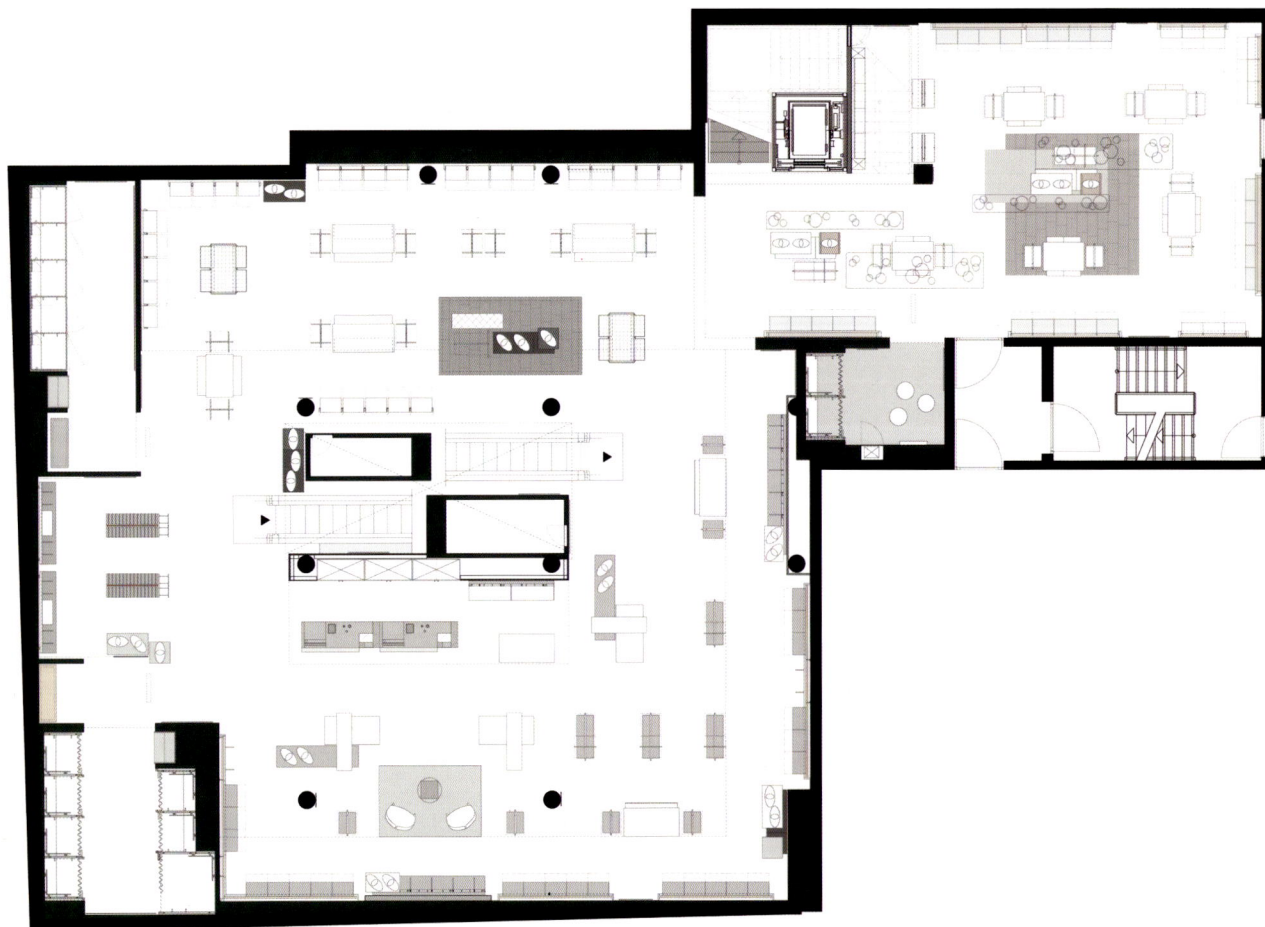

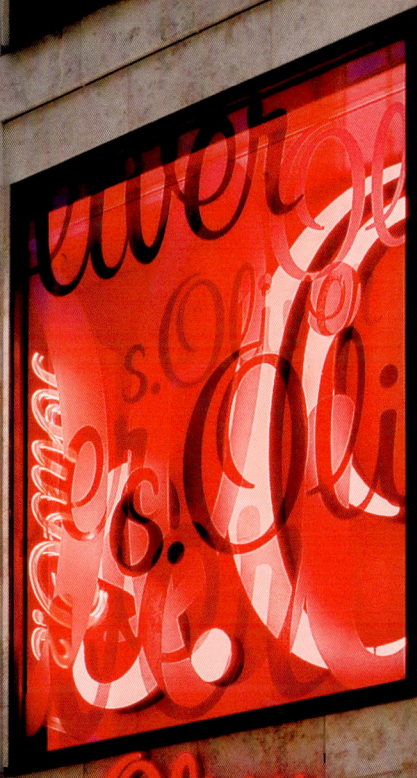

作为欧洲最成功的时装生活品牌之一，s.Oliver满足了所有年龄群的需求，品牌覆盖所有时尚风格，广泛的时尚系列包括"休闲"、"百搭"及"精选"。s.Oliver休闲系列的特点是自然且有质感，配以暖色调色板，营造出舒适、安逸的气氛。

钢铁、深色、开放的天花板将城市年轻人定义为"百搭"的目标群体。专门设计的壁纸配上图形元素，改变了木质饰面的设计，展现出明显的分区，同时也适合老年人群体。

"s.Oliver精选"构成了高端和优雅的商务装。金属饰面与黑白表面相结合。特别设计的金属壁纸强调了精美无光饰面与高光奢华元素之间的对比，如由镜面和黑色玻璃制成的亮点墙。

"s.Oliver少年"适合0～14岁之间的群体。引用了休闲区使用的设计元素，并重新定义了新的光洁饰面。设计师充分利用砖墙面、清新俏皮的元素，如为婴儿设计的花朵和亮色家等，这些都是青少年区域的特点。

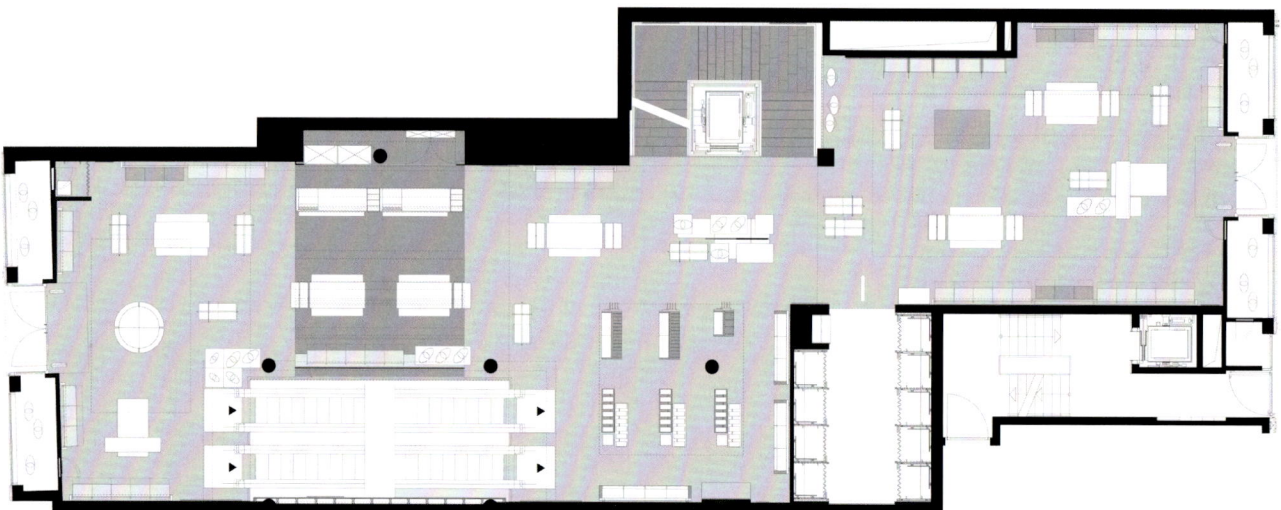

地点	面积	设计公司	摄 影
/ 柏 林	/ 87m²	/ plajer & franz studio	/ diephotodesigner.de ken, schluchtmann

Levis Icon Store Buttenheim Shop

Levis布滕汉姆专卖店

The Levis icon store in Berlin is a small concept store designed by plajer & franz studio. The challenge was to redesign and rebuild the existing store in just 5 weeks before the Berlin fashion week. On top: budget was tight! The vision was to create a feeling of being in a typical old loft apartment with an industrial charm. Historical windows form shelf structures and window bars become cloth rails. Some glass panes are original and others are glazed with acrylic panes with Berlin"places-to-be" printed on.

Old radiators provide a basis for a presentation table which forms the center of the store and allows a varied presentation of products. At the same time it connects the two rooms separated by 2 steps. The changing rooms and the cash desk were built with historic door elements.

Levis merges with Berlin culture by using historic architectural elements in a quite artful way!

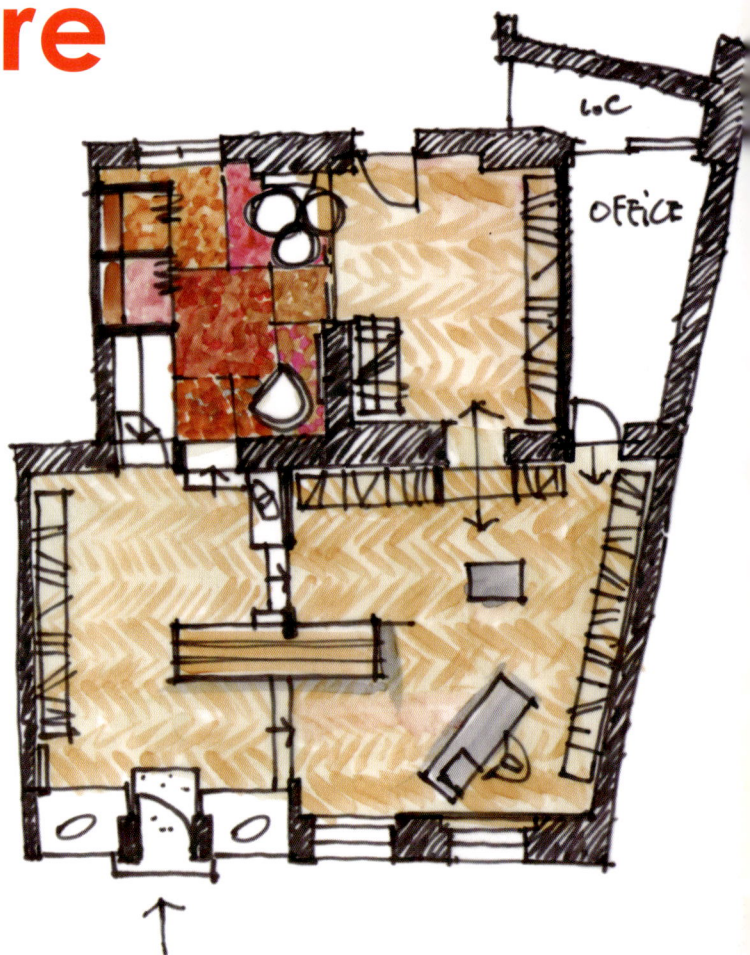

Levis柏林专卖店是一个小型概念店，由plajer & franz工作室设计。设计师面临的挑战是在柏林时装周开始前的短短五个星期内重新设计并重建现有的商店。最棘手的是：预算很紧！

设计师希望营造一种身处典型旧阁楼公寓的感觉，充满着工业魅力。古老的窗户形成搁板结构，窗棂被设计成挂衣服的栏杆。设计保留了一些原有的玻璃窗格，而其他玻璃窗格则被设计为釉面，印有柏林"适合的地方"的字样。

旧散热器提供了基本的展示台，形成了专卖店的中心，可供各种不同产品进行展示。同时，展示台连接着被两部楼梯分割开的两个房间。古老的门户元素被构建成更衣室和收银台。

设计师通过十分巧妙的方式来运用历史悠久的建筑元素，使Levis概念店融合了柏林文化！

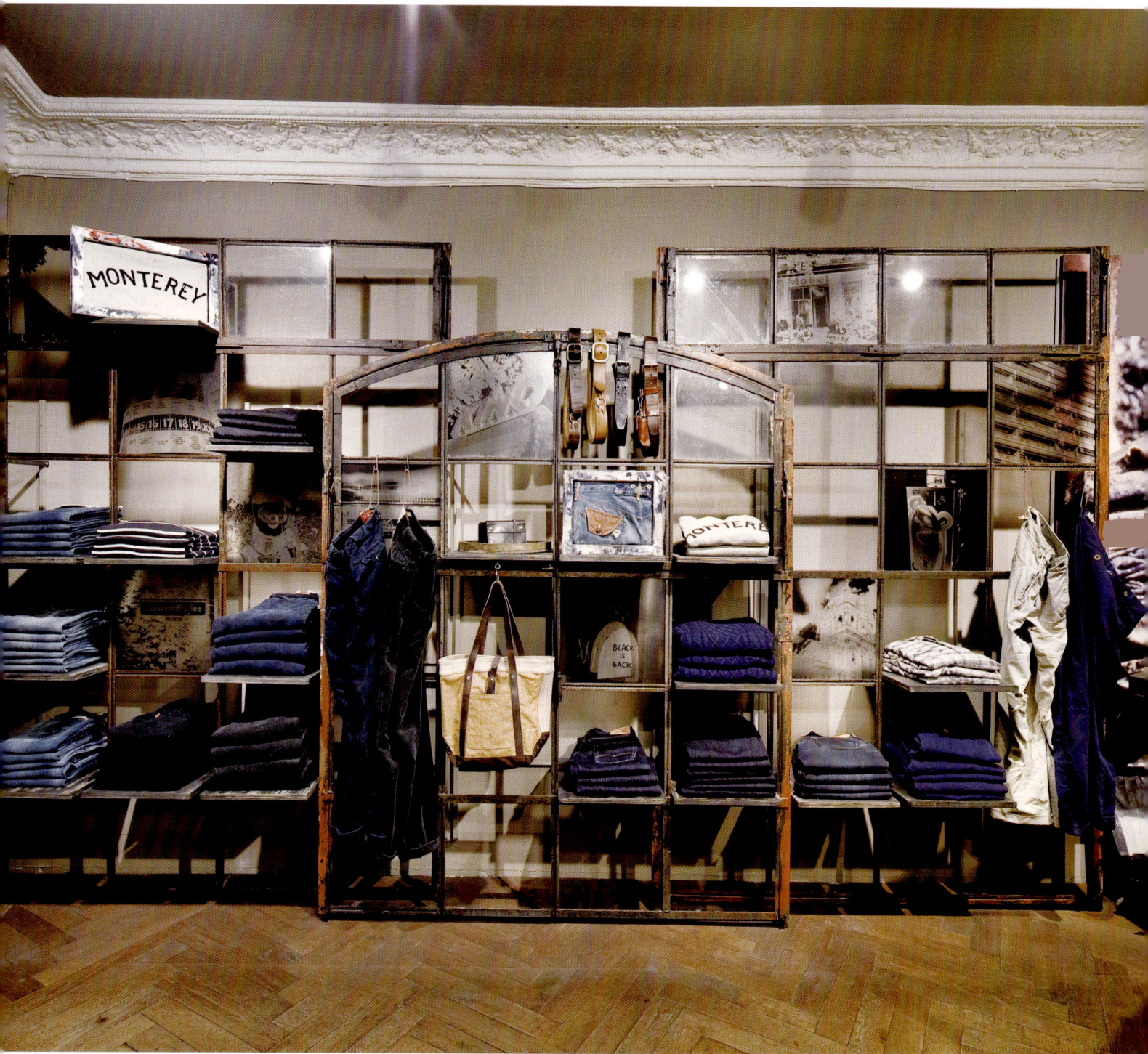

s.Oliver Store Munich

s.Oliver慕尼黑专卖店

s.Oliver has set the course for the future with every aspect of its innovative retail design. The new concept was developed in alignment with s.Oliver's new marketing strategy, with the aim to further strengthen the brand's presence on the market, and was first implemented in March of 2006 in the s.Oliver store in Cologne. While developing the new shop fitting systems, a closer look was taken at the brand's history: special emphasis was placed on the evolution of the s.Oliver parent brand and on the orientation of the individual product segments. Selected materials and specific retail requirements were essential elements, while the needs of

the 25-45 year old s.Oliver clientele, determined in detailed target group studies, represented another crucial factor in developing the new concept. The result is a retail system that reflects an understanding of the brand, its inspiration and experience.

Each s.Oliver store, like the mega store in Munich farbergraben, is a showcase for the trademark. Especially the mega stores are "communicative cornerstones" between the enterprise and the customer, where the brand is represented at its best. Additional special elements are developed for each of these stores.

s.Oliver已经确定了其未来创新零售设计的各个方面。新概念的发展与 s.Oliver新营销策略保持一致，以进一步加强品牌的市场存在性为目标，新概念最早于2006年3月在s.Oliver科隆专卖店实施。

在开发新店面装修系统的同时，设计师更注重品牌的历史：特别强调s.Oliver母品牌的演变和个别产品分类的方向。所选的材料和具体零售要求是必不可少的元素，而详细目标客户研究发现的25~45岁s.Oliver客户群的需求，则代表着新概念发展中的另一个关键因素。结果是零售系统应反映对一个品牌的了解以及它的灵感和经验。

每个s.Oliver专卖店都是商标的展示，如慕尼黑大型专卖店。尤其特别的是，大型专卖店是企业与客户之间的"沟通基石"，在这里，品牌是最好的代表。各个专卖店还额外设计了其特殊元素。

Tangy Collection_
Shenyang

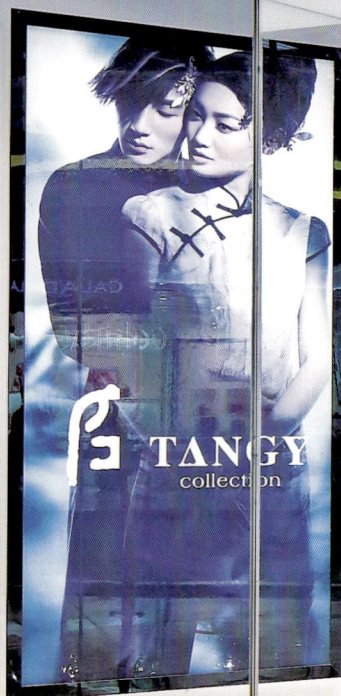

—— 沈阳天意

TANGY Collection is designed with the concept of "peace, health and beauty" in the pursuit of the harmonious state of "integration of man and nature". As the design of its brand store, the style of the store should be combined with the concept of the clothing brand, and the design of the store just follows its essence. Adhering to the concept of "harmony between man and nature", the interior design is in great integration with the brand: comfortable and low-key. The whole space is simple and fresh, where the men and women clothing are shown separately, but interrelated with each other. Prominent square and spotlights on the ceiling shine well with each other, endowing the simple space with a touch of charm and adding a little more personality into fashion.

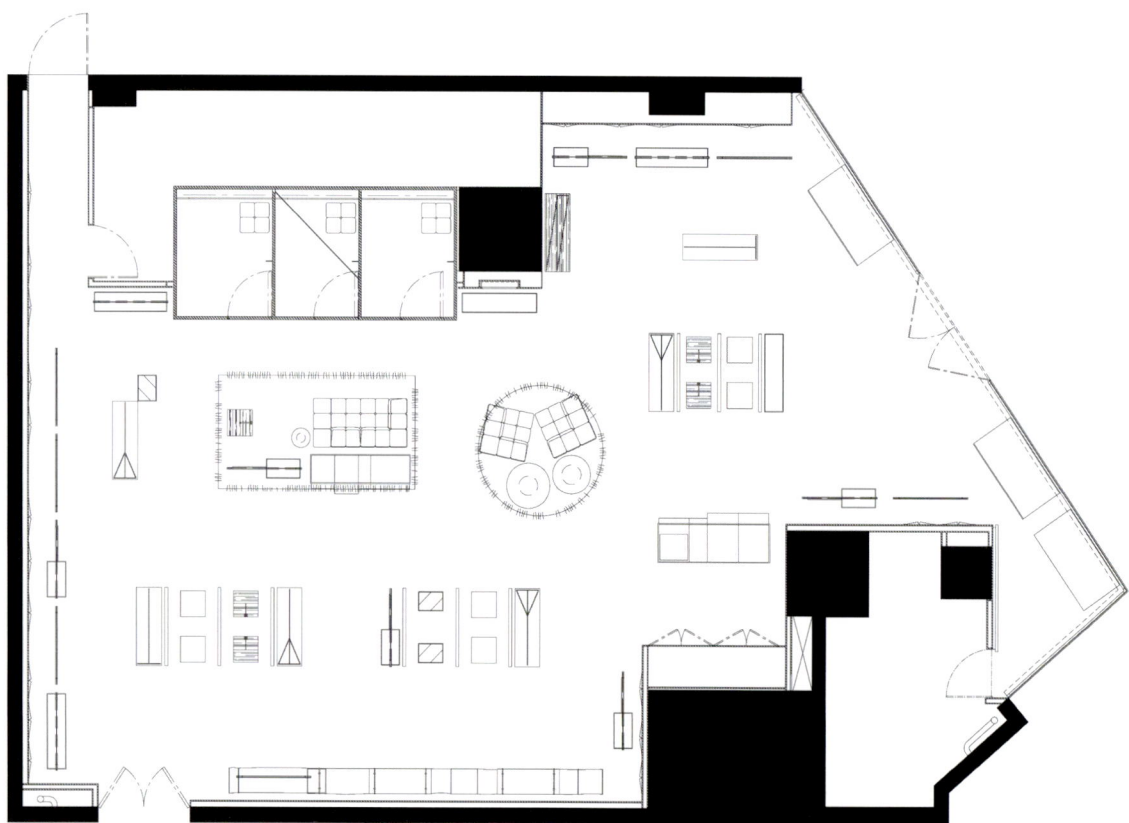

TANGY
collection

　　沈阳天意专卖店以"平和、健康、美丽"为理念，追求"天人合一"的和谐境界。作为其品牌店的设计，店面风格应与服装品牌的理念合二为一，此店的设计可谓是得此真髓。秉承"天人合一"的观念，室内设计与品牌很好地融为一体：熨帖而低调。整个空间简约而清爽，男女服饰分区展示，却又相互关联。突出的四角形与天花板上的射灯辉映成趣，为简约的空间平添了几分意趣，也让时尚中多出了几分个性。

TANGY
collection

Tangy In Shenzhen
—— 深圳天意

"TANGY" is a clothing brand conveying the concept of nature and human coexistence, and the design is inspired by the designer's unique thinking of the brand.

The characteristic of "TANGY" brand is to show the traditional Chinese dyeing technology and unique feel of modern textile on the plain cloth. What the designer wants to show is the dyed materials, and the theme network structure mainly with cross section of plant cell holds the whole space. The ninety four 2000×1200mm panels form a smooth and continuous network, covering the ceiling and walls, so the whole space is like an art gallery. The mirror aluminum ceiling and swing furniture present the artificial materials naturally.

"天意"是传达自然与人类共存的理念的服装品牌，其设计灵感来源于设计师对于该品牌的独特思考。

"天意"这个品牌的特点是将中国传统的染色技术与现代纺织的独特手感体现在自然素材布上。设计师要表现的则是染色原料，以植物的细胞断面为主题的网状结构包囊了整个空间。94张2000mm×1200mm的面板组成光滑、连续的网，将天花板和墙壁包覆住，让整个空间犹如一个艺术展厅。镜面铝的天花板和摇摆不定的家具，将人工素材自然地呈现。

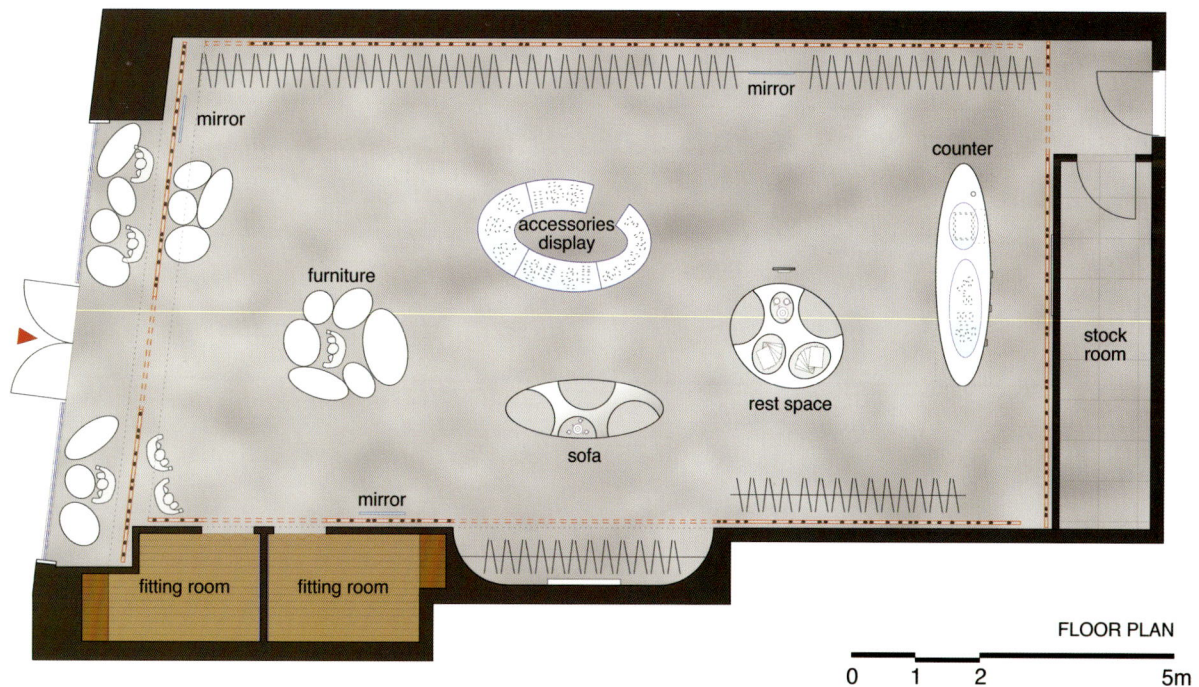

mirror

mirror

counter

furniture

accessories
display

stock
room

rest space

sofa

mirror

fitting room

fitting room

FLOOR PLAN

0   1   2   5m

Scfashion Shenyang

—— 沈阳Scfashion

This is an exclusive women's clothing store. The space is mainly simple and stylish, creating a fashionable and capable brand store for urban ladies. The space is divided into three parts in the plan: the products display area at the entrance, the photography area or rest area in the middle, and the warehouse and changing rooms in the back. In the building of the facade, the designers introduce clothes hangers interspersed with hanging curtain and wood finishes, making the moving lines in the space clear and neat. The ceiling is designed with veneer interspersed with spot lights, giving judicious guidance according to circumstances, so that the passengers' movement is simpler and more straightforward. The control of colors takes the original colors composition of women's clothing into account, and the space is designed with white combined with beige to create an elegant and generous temperament, interspersed with the color orange to brighten the space, making the space more spiritual.

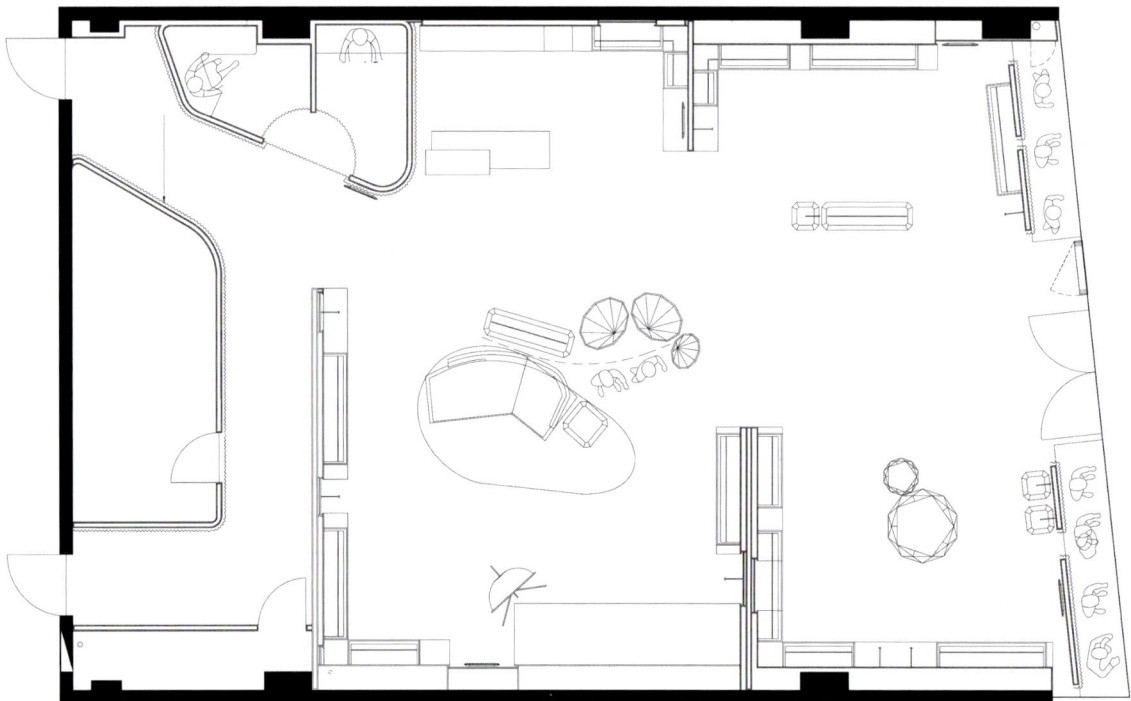

这是一家专营女装的服装专卖店。空间以简约、时尚为主线，为都市女性打造出一个时尚、干练的品牌店。空间在平面上分为三个部分：进门处的产品展示区、中间的摄影区或者休息区、最里面的仓库和试衣间。在立面的打造上，设计师利用挂衣架穿插挂帘、木饰面，让空间动线明晰、利落。天花板的设计上，利用饰面穿插结合射灯分布，因势利导，让客流动向更简单明了。在色彩的控制上，将女装本身的色彩构成考虑进来，空间运用白色结合米色营造出素雅大方的气质，穿插橙色提亮空间，也让空间更有精神。

Tangy Collection _Tianjin

—— 天津天意

TANGY COLLECTION in Tianjin is small, and the customer can see all from the outside. Thus, the designers add a curtain behind the display area with glass doors, and looming is more attractive. The black "TANGY COLLECTION" is not only the sign, but also a direction to guide the customer to come in from here.

The interior design is in accordance with simple and stylish line, where the simple and clear moving lines for display are arranged, and suspended hangers is combined with fold-line showcase which is full of dynamic lines to enrich the space form. The curtain-like dense slings and the prominent quadrangular stars on the wall are added into the space, slightly showing softness from toughness, perfectly combing the space with the brand.

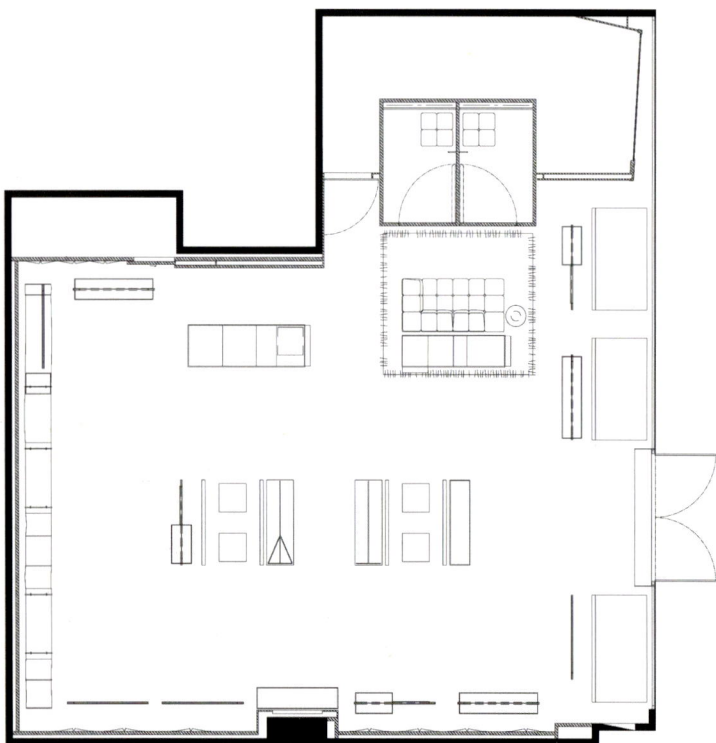

　　这是"天意"在天津的店，店面面积不大，从外面一眼便可看穿所有。于是，设计师在玻璃门的展示区后加了一层隔帘，若隐若现才更能引人入胜。黑色的"天意"既是招牌，又是一种指示，指引顾客从此处进入。

　　店内的设计依照简约、时尚的路线，安排了简单明了的展示动线，悬吊式的挂衣架结合富有动感的折线展示架，丰富了空间形态。加上细密如帘的吊索与墙面突起的四角星，干练中隐现丝丝温柔，将空间与品牌很好地融合在一起。

TP Internationally Famous Flagship Store

TP国际名品旗舰店

TP internationally famous flagship store is in prosperous area at the city center. The designer takes advantage of the characteristics of the original high structure, ensuring the generous and high momentum of the lobby, and the latter part of the store is divided into two layers by steel structure with the men's clothing on the first layer and the women's clothing on the second layer. It enriches the beauty of the space, and increases the utilization of the space as well. However, this design makes the second layer comparatively low. So in the design of suspended ceiling, the designer chooses exposing style to retain the original beam structure, beautifying the space and ensuring the height.

Another feature of the project is the use of mirror reflection. No matter the silver mirror finish on the floor and the black mirror on the background wall, or the floor-to-ceiling mirror in the hallway at the second layer and mirror stainless steel decoration in details, they fully show the extension of the space and the unique atmosphere of fashion store.

In the color application, the project is mainly black, white and gray. The large area of black mirror, French off-white stone with wood texture coupled with gray vitrified tiles, and generous white ceiling of soft film create a charming display space for the diversified fashion clothing.

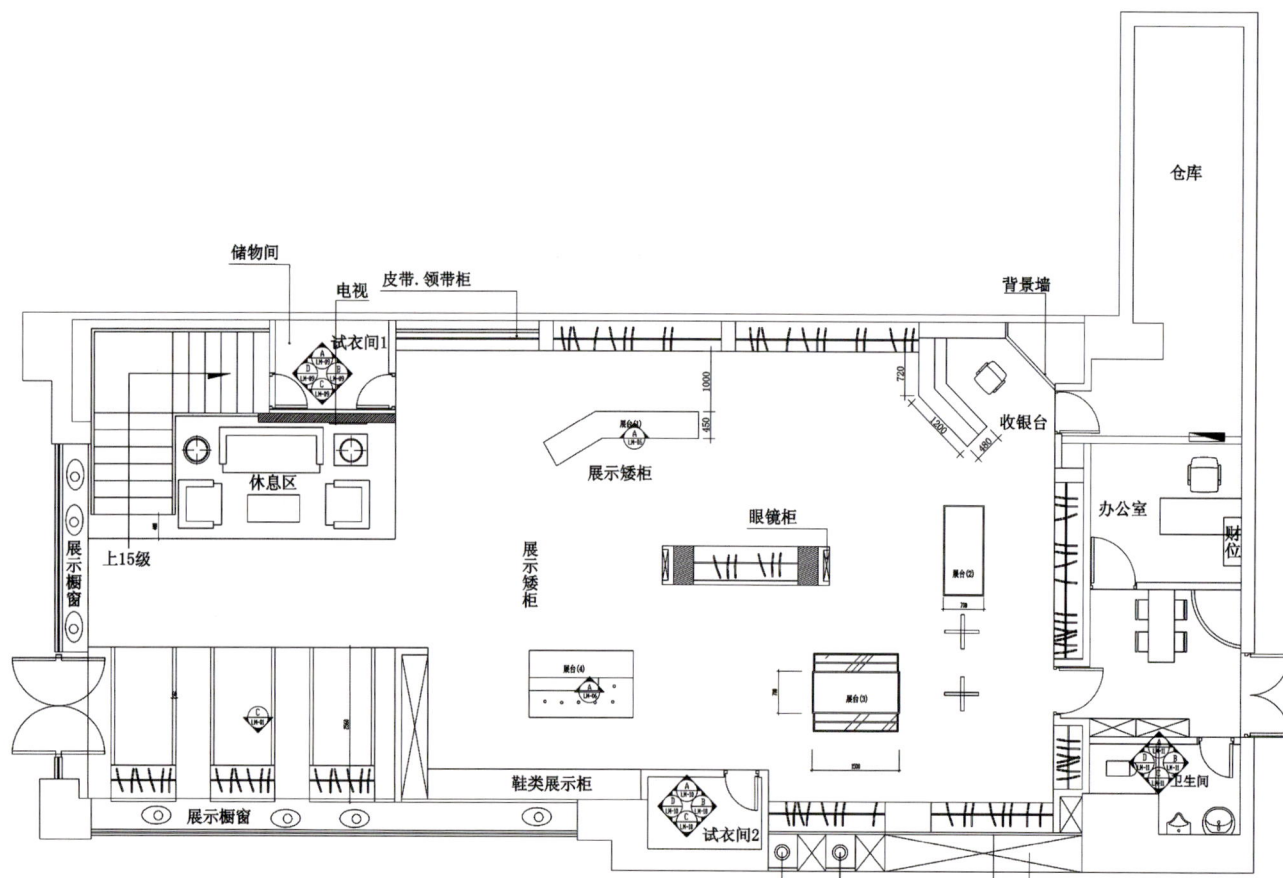

　　TP国际名品旗舰店位于城市中心繁华地段。设计师利用了原有结构楼层高的特点，在保证前厅大气、高挑的气势后，在店铺的后半区，用钢结构隔层划分出一层男装区域及二层女装区域，既丰富了空间美感，又增加了空间利用率。但是，此种设计使二层的层高偏低，所以设计师在吊顶的处理上，以暴露式设计为主，保留原结构梁的构造，既美化了空间，又保证了高度。

　　本案的另一特点是利用镜面反射。不管是造型门的银镜饰面，背景墙黑镜的应用，还是二层过道的落地镜及细节处的镜面不锈钢装饰，都充分体现了空间延伸感及时装卖场特有的气息。

　　在色彩应用上，本案以黑、白、灰为基调。大面积的黑镜，米白的法国木纹石搭配灰色玻化砖，以及大气的白色软膜天花都为多样化的时装创造了一个迷人的展示空间。

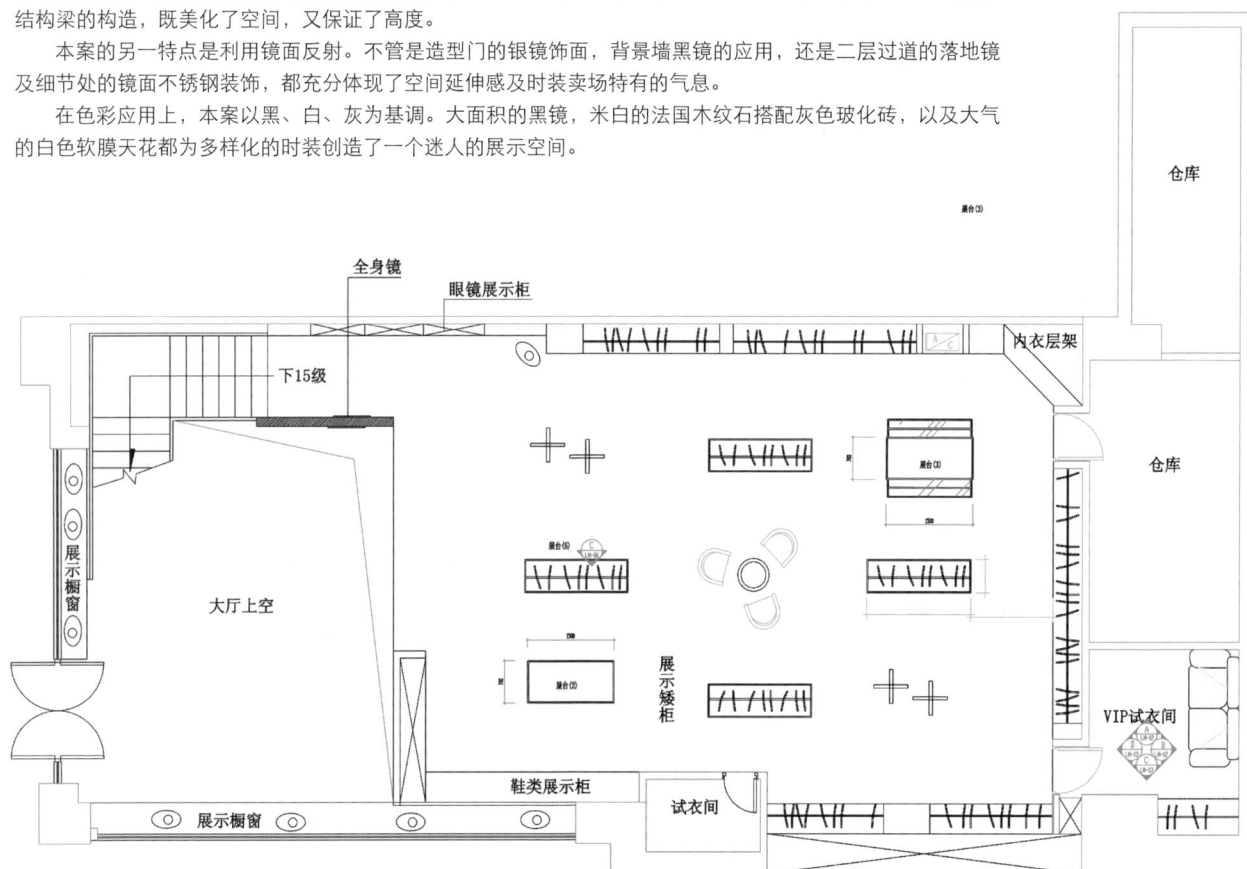

D&G
DOLCE&GABBANA

PRADA

DSQUARED²

FERRE

HALIL & JOE

PATRIZIA PEPE
FIRENZE

BIKKEMBERGS

DOLCE&GABBANA

地点	面积	设计师	设计公司
/ 汕头	/ 143m²	/ 李伟光	/ 汕头市丽景装饰设计有限公司

Da Ying Fashion

大迎服饰

The store mainly sells men's clothing and related accessories. So the design is simple and capable, fashionable and tough style, coupled with clear mirror, black iron and other cold and hard materials to create a handsome and masculine space for men. The structure is in "L" shape, and the interior layout is simple, bright and clean. Starting from the facade and ceiling, the designer uses undulate stepping, coupled with black iron hangers, to make the interior rich and tough. The ceiling is combined with the original structure of the space, integrated with clear mirror to form strip-shaped guide groove to lead the moving line, making the space more active and colorful. In the aspect of color, white is the main color coupled with copper metal, interspersed with black iron frames, especially cool.

本店主营男性服装及相关饰物，因此店面的设计以简单干练、时尚硬朗为风格，配以明镜、黑铁等冷硬的材料，打造出一个阳刚帅气的男性空间。

本店的结构呈"L"形，空间内部简单、明净。设计师从立面和天花着手，立面用高低起伏的垫脚配合黑铁挂衣架，让里面显得丰富而硬朗。天花则结合空间原本的结构，利用明镜做出条状导引槽，引领人流动线，让空间更为活跃丰满。色彩方面，白色主调搭配铜色金属，再穿插黑铁构架，帅气十足。

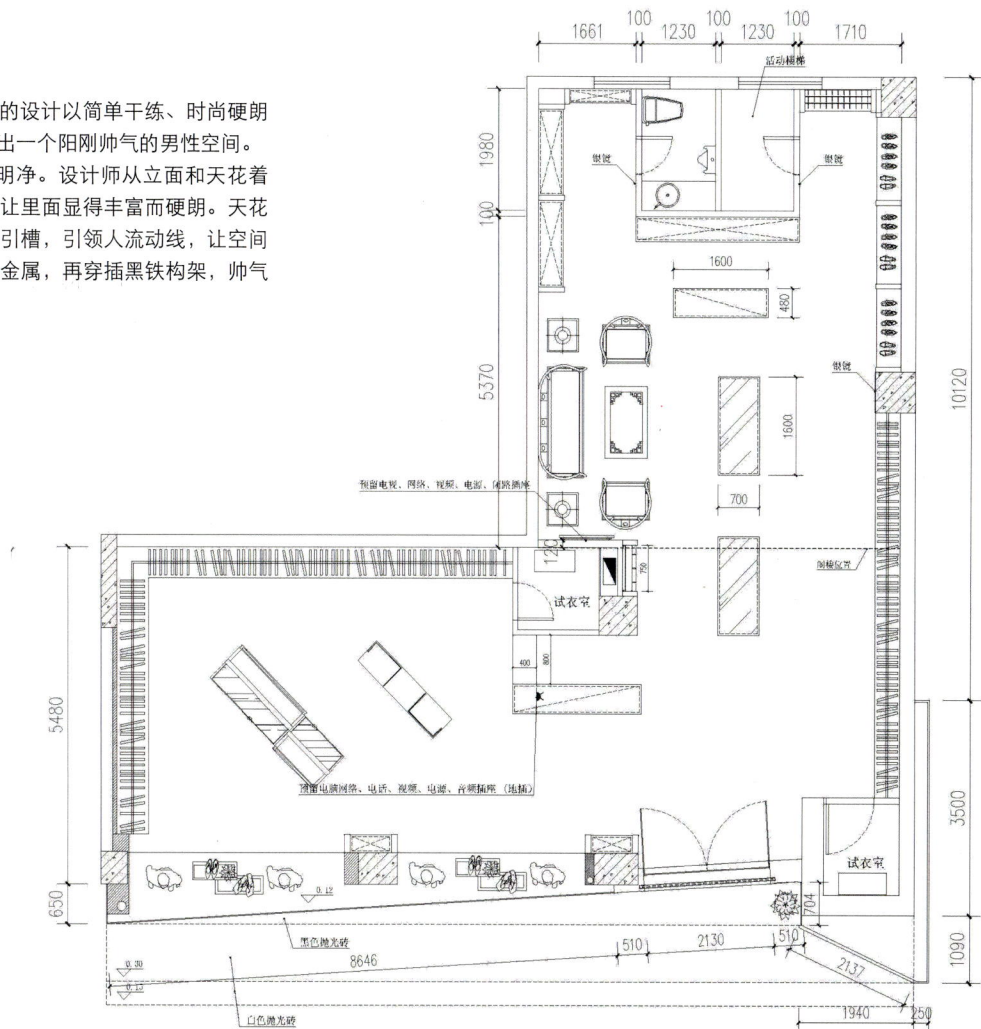

Men Today
今日男仕

As a clothing store mainly selling men's accessories, Men Today combines the sales of men's clothing, leather goods, shoes and hats, glasses, bags and other items. On that basis, the primary task of the store design is to limit the functional partition, followed by the definition of design style.

In the store, we can see a wide range of businesses, but in order, and the simple and refreshing layout is coupled with colors collocation in sharp contrast of white and black, allowing the whole space to show a fresh and agile feeling. Such kind of design is the result of integration of operating project and consumer group, and with the premise of meeting consumers' desire, being simple and fashionable is the real caring and humanization.

今日男仕作为一家主营男士饰物的服饰店，综合了男士服装、皮具、鞋帽、眼镜、箱包等多种物品的销售。在此基础上，店面设计的首要任务是功能分区的限定，其次才是设计风格的界定。

在这个店里，经营的种类繁多，但是却一点也不显得繁乱，简洁清爽的布局加上黑白分明的色彩搭配，让整个空间呈现出一种清爽利索的感觉。这样的设计是综合了经营项目和消费人群特征之后的结果，在满足消费者意愿的前提下，做到的简约时尚，才是真正的贴心和人性化。

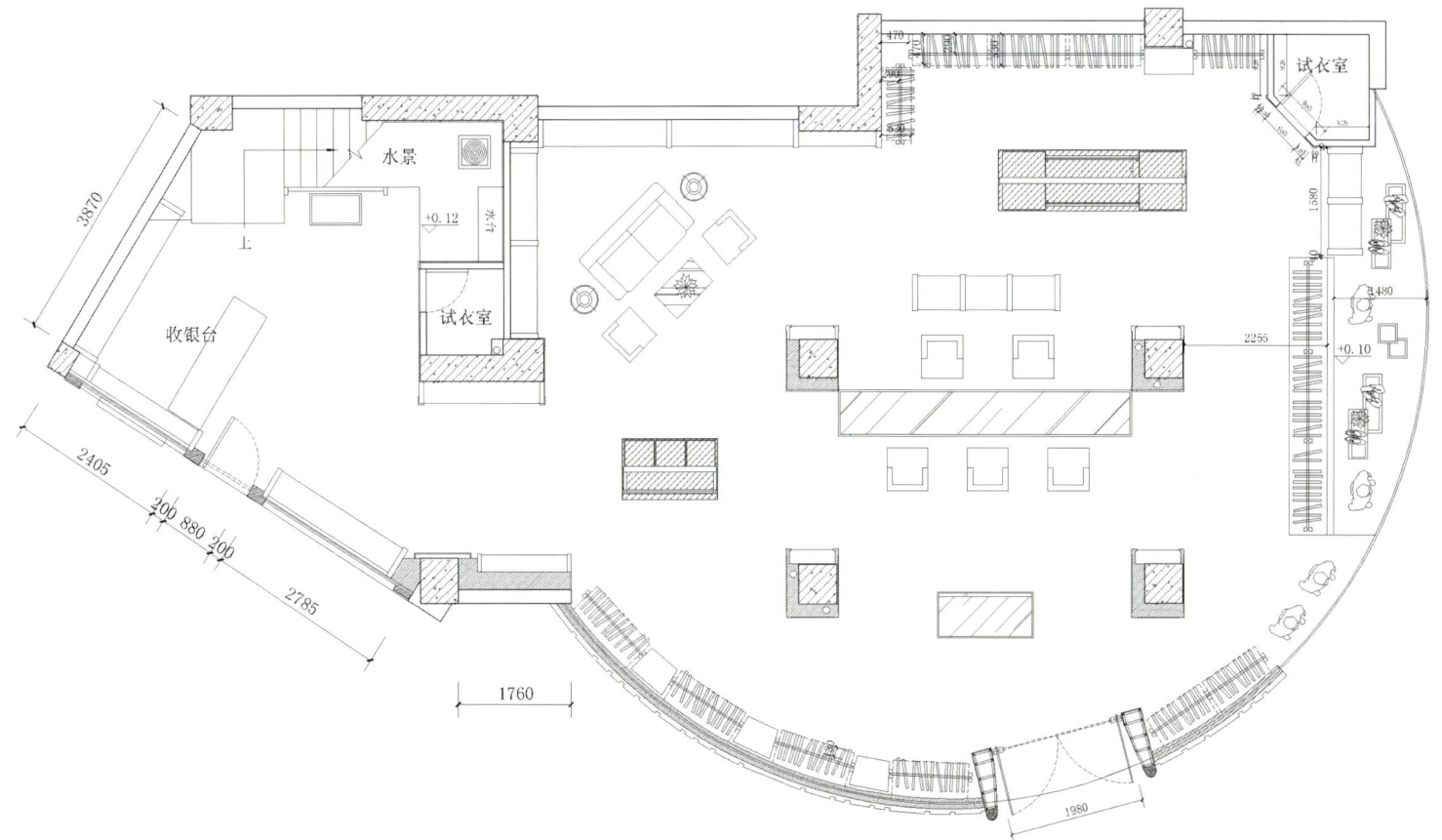

ACCESS-
ORY 配饰

GARA Boutique   GARA精品店

Oekonia Jewelry Store of Italy   意大利欧卡拉饰品店

American Propet   美商波派

D.A.BEAU in Shenzhen CITIC Seibu Department Store   D．A．BEAU深圳中信西武百货

Conradt Optik   Conradt Optik眼镜店

Dovinie   德维尼

Nine Silver Mill   九银坊

GARA Boutique

___ GARA精品店

The world famous GARA boutique is located in the beautiful Outlets Shopping Village in Shenzhen Da Meisha, is a discount shop selling GUCCI, PRADA, MIUMIU, DIOR, FEDI, HERMES, and other world's top fashion merchandise.

The designer takes flower as inspiration, extending to the vase, petal, bud and other designing elements, and introduces the flawless pure white color and black color background for combination to create a spotless shopping environment with elegance, and to provide a harmonious and united exhibition and retail space for various unique famous fashion brands.

BOSS

GARA世界名牌精品店位于美丽的深圳大梅沙奥特莱斯购物村,是一间销售GUCCI、PRADA、MIUMIU、DIOR、FEDI、HERMES等世界顶尖时尚精品的折扣店。

设计师以花为设计灵感,延伸出花瓶、花瓣、花蕾等设计元素。配合无瑕的纯白色调和黑色的大背景,营造出一尘不染的高贵气质的购物环境,同时也为独具个性的各大时尚精品品牌提供了和谐统一的展示、零售空间。

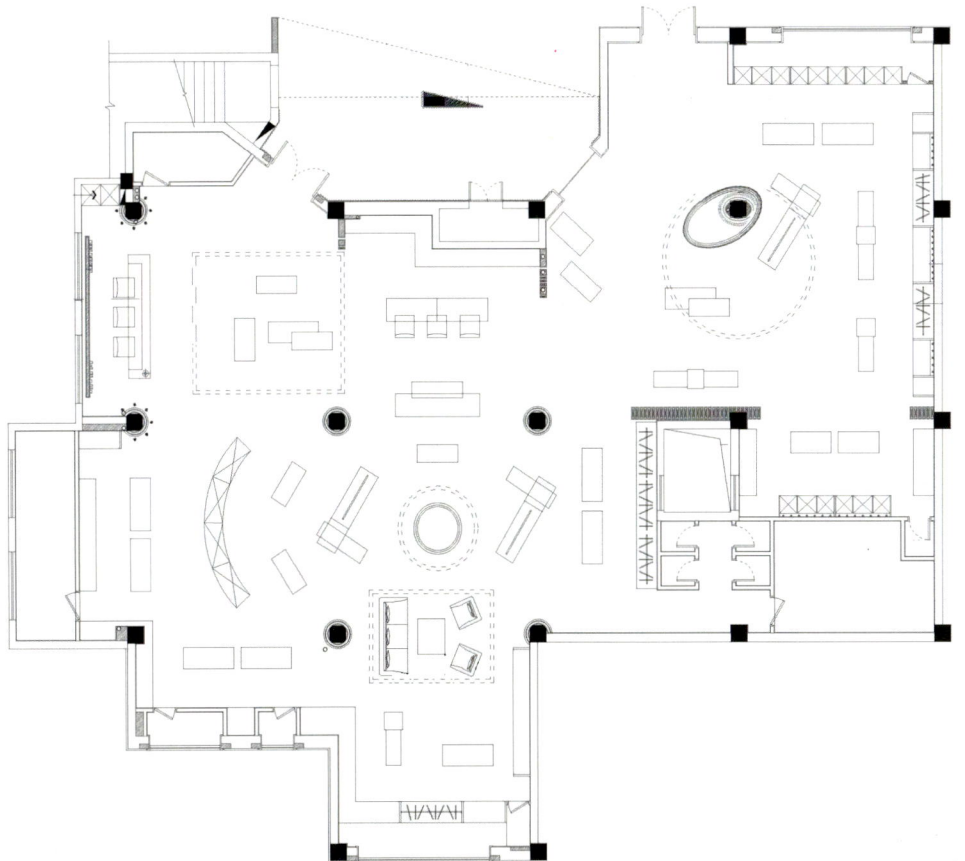

地 点	面 积	设计师	设计公司		主要材料
/ 宁 波	/ 60m²	/ 张向东	/ HBS+宁波红宝石装饰设计有限公司		/ 镜面不锈钢、墙纸、玻璃、复古砖

Oekonia Jewelry Store of Italy

意大利欧卡拉饰品店

Oekonia is an internationally known brand, the first domestic exclusive shop of which is settled down in Ningbo. How to express the crystal clear and colorful product expressions of colored glaze decorations in the small space is the main problem of the design.

The designer takes the target customer group, the women pursuing personality and fashion, as the breakthrough point. He is inspired by the fun of kaleidoscope which can change into gorgeous and magic images through several pieces of glass of different colors. Therefore, this small space reflects a kind of magic which makes it full of illusion, fun and practical applicability.

The interior design takes the rose flower as the theme of the background and uses the LED seven-color scanner light as the auxiliary, which enriches the color of the whole space and brings the visual illusion and dynamic as well. Stainless steel of the whole wall space and mirror face of the top surface reflect and form another subjunctive and colorful world. The size and height of two rows of floating "display diamonds" are totally different, which forms strong visual appeal to the customers. The lighting within the diamonds reflects up and down, the small and exquisite colored glaze jewelry is glittering and translucent.

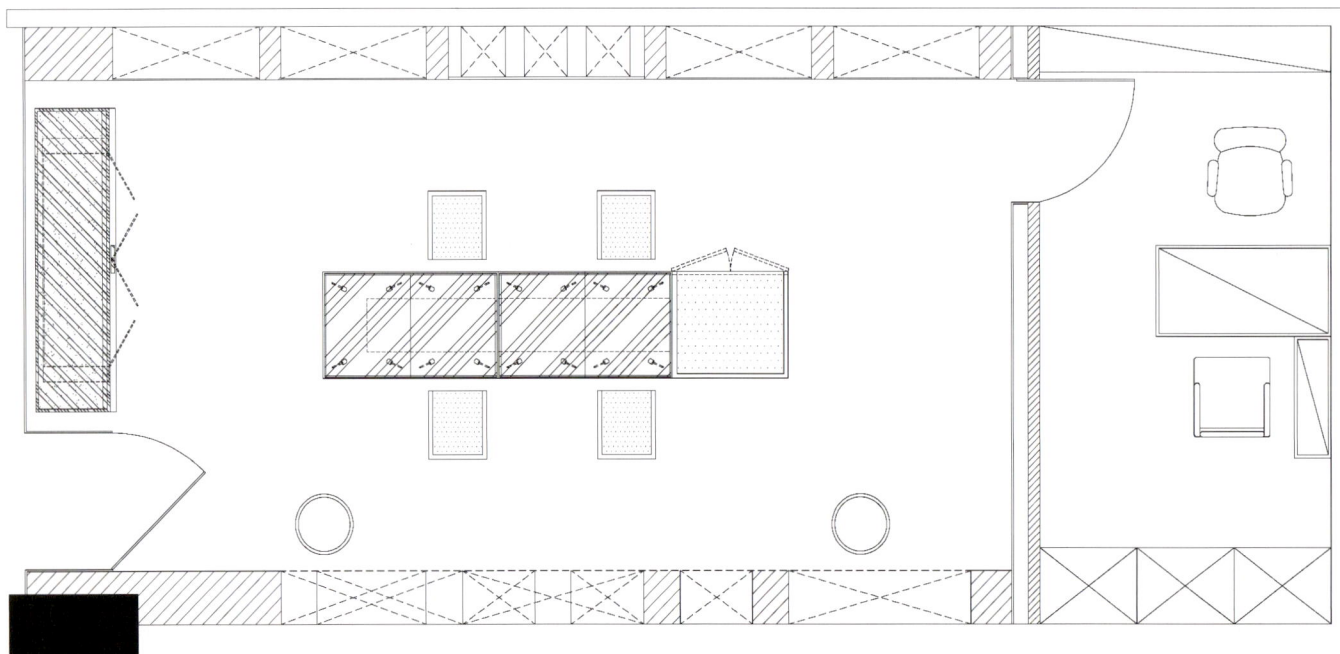

　　国际知名品牌欧卡拉的国内首家专卖店落户于宁波,如何表达出琉璃饰品晶莹剔透、五彩斑斓的产品表情,是设计的首要问题。

　　设计师从产品的目标客户群——追求个性化、时尚化的女性——为切入点,以万花筒的趣味性——通过几片不同颜色的小玻璃片变幻出绚丽、神奇的图像为灵感源泉。因此,这个小空间是富有迷幻感、趣味性和实用性的。

　　室内设计背景以玫瑰花为主题,并辅以LED七色扫描灯,丰富整个空间色彩的同时,也带来了视觉的梦幻和动感。整面墙面及顶面镜面的不锈钢也折射出一个虚拟的缤纷世界,两排浮动的"陈列方块"大小、高度各有不同,对顾客有强烈的视觉吸引力。方块内灯光上下辉映,小巧精致的琉璃首饰,晶莹剔透。

地 点	面 积	设计师	设计公司	主要材料
/ 广 州	/ 800m²	/ 云男、黄世平	/ 深圳三迪设计顾问有限公司	/ 抛光砖、大理石、饰面板、皮革

American Propet

—— 美商波派

American Propet Group is an image store of new-style shoes, located in the first floor of the office building. The functional partition meets customers' demands, where the whole design is mainly white, simple, elegant and bright, and the round-in-square elements are introduced throughout the design.

The designers start from two points: the "common character" and "individuality" of the space. Starting from the construction, the common character takes the relationship between architecture and space into account to decide the space sequence, so the architectural form, light and materials become the first issue to be settled. The designers are inspired by the architecture and decide to begin with "arc" to the end. The roof changes into various lines of arcs featuring modern and historical sense, allowing the spirit to leap in the space freely. After the completion of the space construction, the "individuality" of the space should be considered. Furniture, furnishings, lighting and other elements are introduced to create a suddenly near or far intent, and these neo-classical elements supplement the "common character "as well, for example, the line on the ground and the one on the ceiling form a kind of complementary unity.

美商波派国际有限公司位于办公大楼的首层，是新款鞋形象店。功能使用分区满足了客户的需求，整体设计简洁、大方、明快，以白色调为主，方中带圆的元素贯穿整个设计。

设计师从两个问题出发：空间的"共性"与"个性"。共性从建筑着手，考虑建筑与空间的关系，决定空间的序列，于是，建筑形态、光、材质等成为其首要解决的问题。设计师从建筑中得到灵感，决定以"弧"开始，以"弧"结束。屋顶变成一条条兼具现代感与历史感的弧线，让精神可以在空间中飞跃，获得自由。在空间构造完成后，便要考虑空间的"个性"问题了。通过注入家具、陈设、灯具等元素，营造出一种忽远忽近的意境，这些新古典主义的元素同时也能补充空间的"共性"，例如，地面的线条与天花的线条，形成了一种互补的统一。

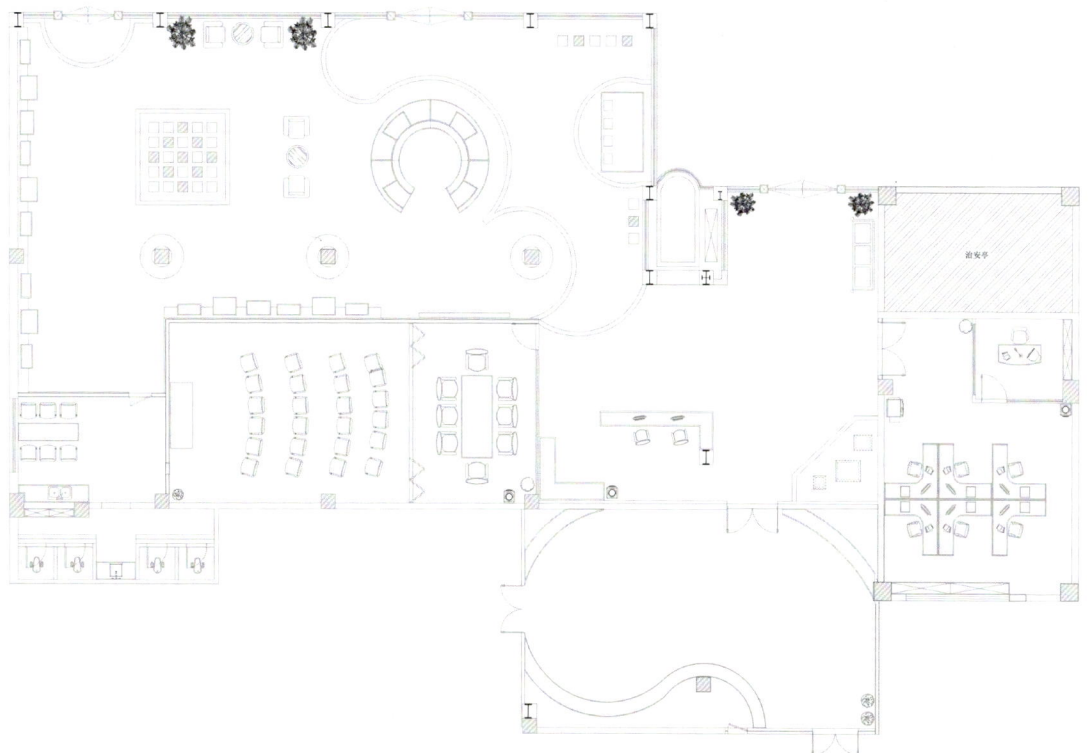

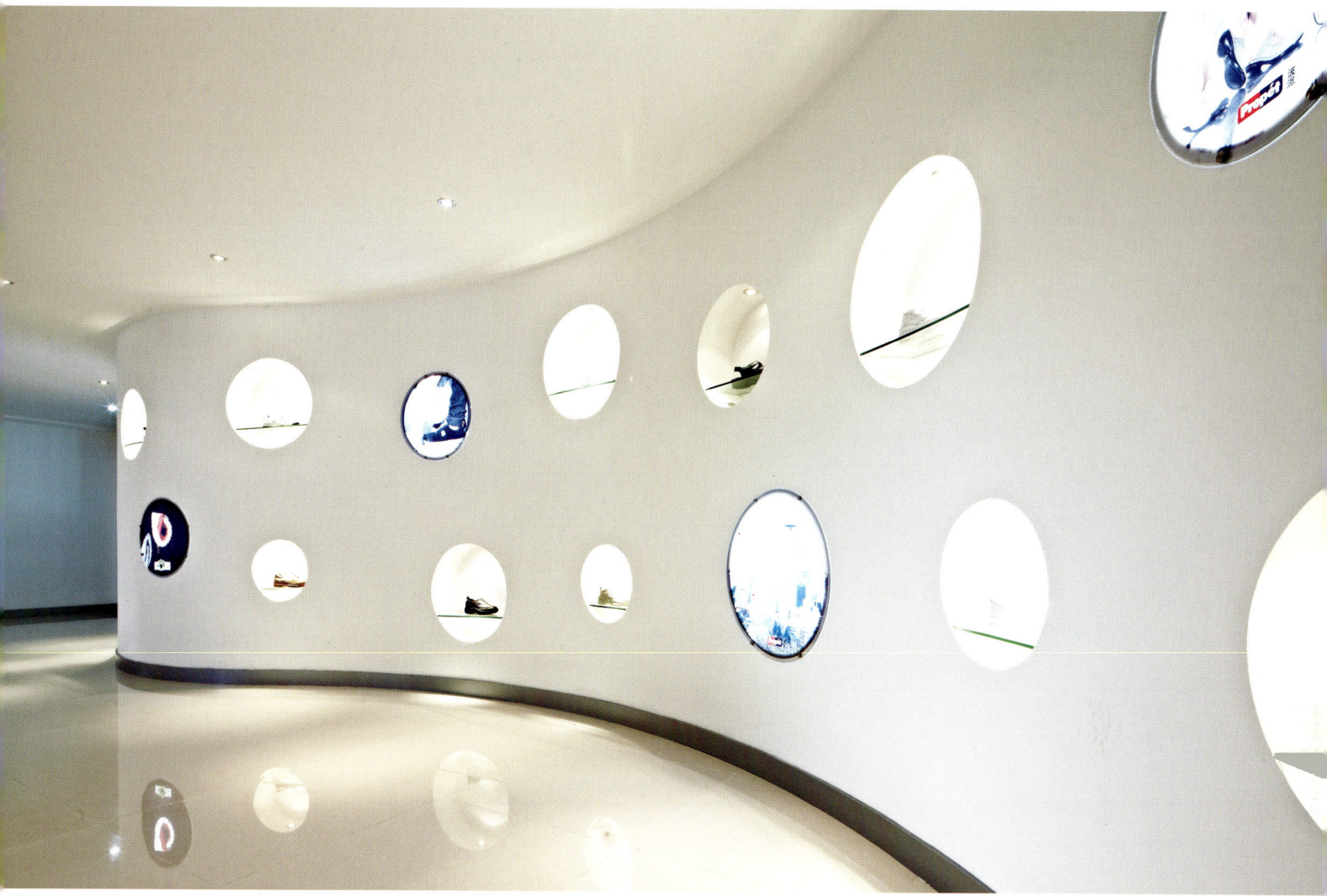

D.A.BEAU in Shenzhen CITIC Seibu Department Store

D.A.BEAU深圳中信西武百货

D.A.BEAU is a bag exclusive shop, and the whole design is simple and fashionable, in line with the business concept of the brand. The design of the facade is simple and practical, and customers can probably see the interior layout from outside and the items on sale and other information. In addition, the facade is designed in a good combination with the fire hydrant, where the design of the hidden handle is coupled with the painting of the brand spokesman to make the wall more three-dimensional.

The interior design is mainly in copper gold and white colors. Where the white wall is embedded with copper gold cabinet, and the copper gold edges divide the ceiling and the selling area, making the space more three-dimensional and more personalized and fashionable. Such kind of conciseness highlights the charm of the space, and reveals the personalized charm of the brand.

消防栓

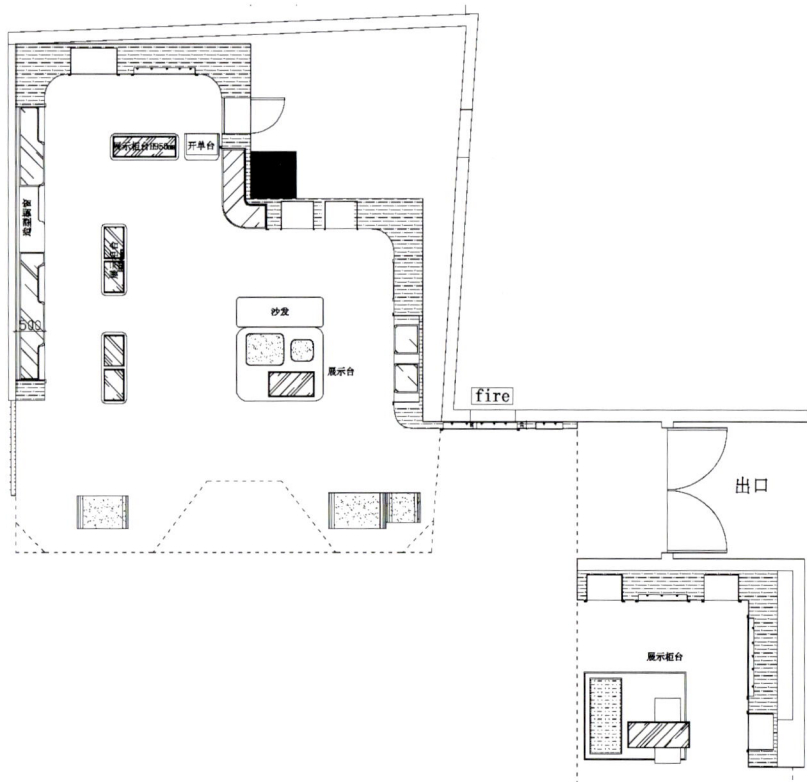

本案为箱包专卖店，整体的设计简约、时尚，这符合了该品牌的经营理念。门面的设计简单却很实用，让人从外面就能大概看到室内的布局及经营的具体项目等信息。此外，门面还很好地结合了消防栓进行设计，隐藏式把手的处理搭配代言品牌的明星挂画让墙面更为立体化。

店内的设计以铜金色和白色为主，白色的墙面镶嵌铜金色的柜架，同时铜金色的边线将天花与售卖区分割开来，让空间更有立体感，同时也显得更为个性时尚。这样的简约既凸显出空间的魅力，又彰显了品牌的个性魅力。

地点	面积	设计团队		设计公司	摄影
/ 德国	/ 210m²	/ Peter Ippolito, Gunter Fleitz, Alexander Fehre, Tim Lessmann, Christian Kirschenmann, Vincent Gabriel, Anne Lambertz, Axel Knapp (Graphics), Yuan Peng (Graphics)		/ Ippolito Fleitz Group	/ Zooey Br

Conradt Optik

Conradt Optik眼镜店

The refurbishment of Conradt Optik coincides with the business being handed over to the next generation of owners, who wish to position the business with a select segment of brands and a focus on individual customer care. The interior can be accessed through one of two entrances, both of which lead to the central service counter. An organically curving rear wall masks the workshop areas and divides the shop into zones. Fitted into the rear wall are long backlit presentation bays and flush-set drawers for accessible product storage. The recessed ceiling is decorated with a pattern of fine lines and differentiates the customer service area from the shop floor.

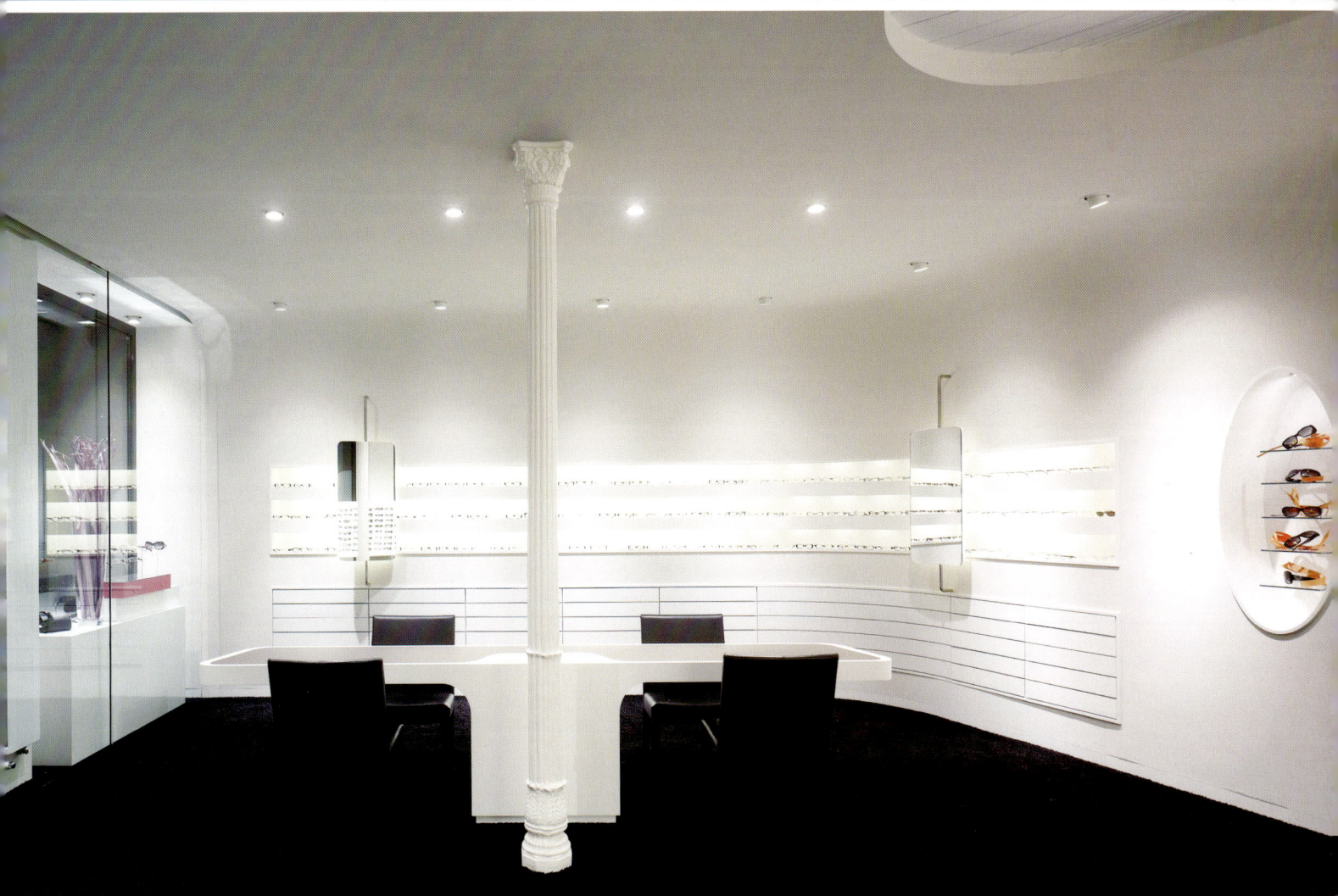

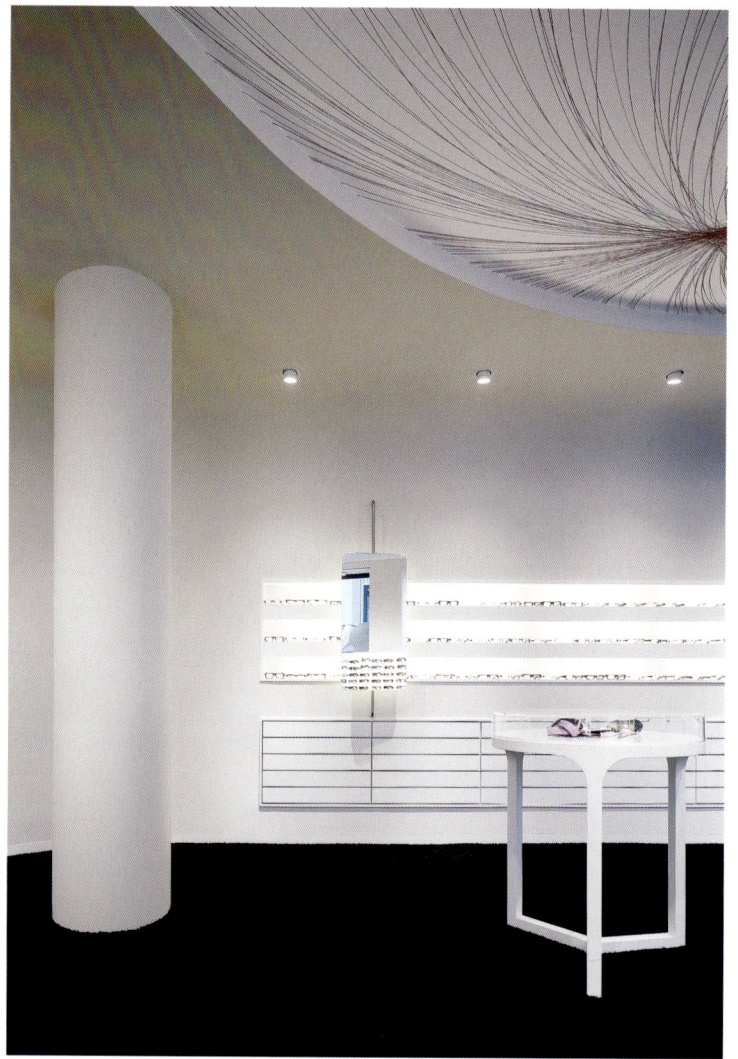

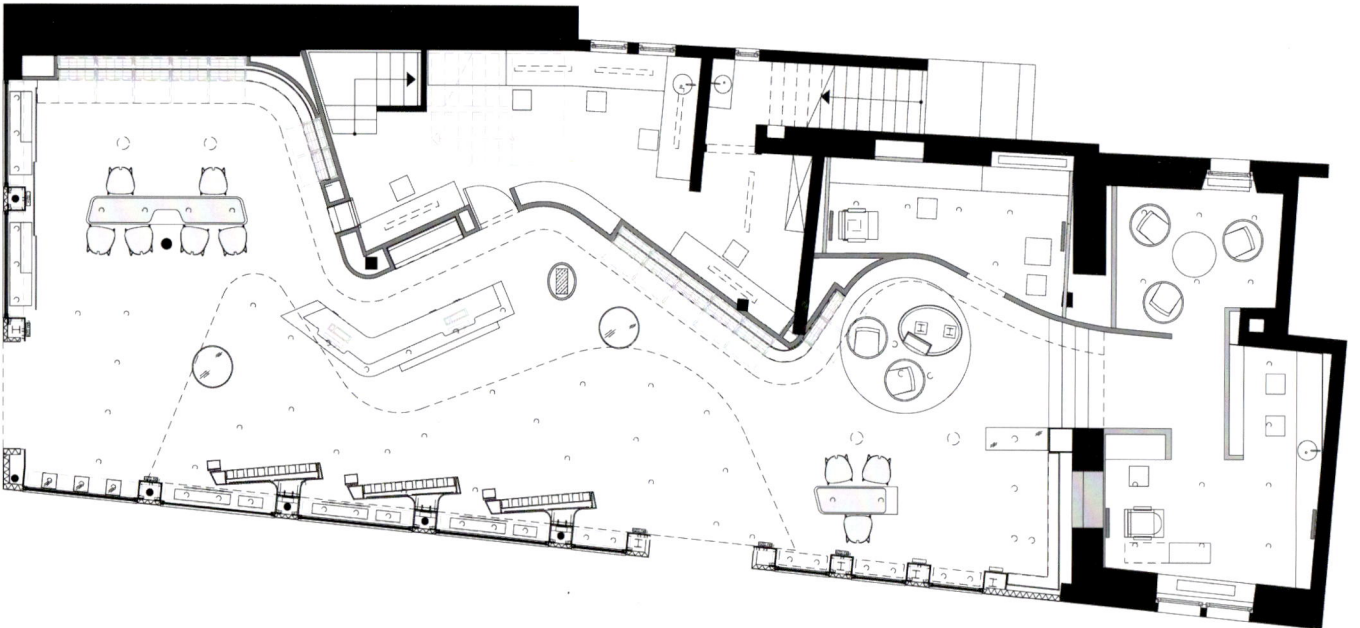

Conradt Optik眼镜店的翻新恰逢下一代业主接手公司业务，业主希望将业务定位为品牌精选、重视个体客户需求。可以通过两个入口中的任意一个进入室内，两个入口都通向中心服务台。有机弯曲的背景墙掩饰了加工区，并将商店分成若干区域。背景墙镶嵌了长长的背光演示带，顶格排列的抽屉可存放眼镜。吊顶饰有精美的条纹图案，将客户服务区与加工区分开。

Dovinie

—— 德维尼

It is a jewelry store, and the interior design requires featuring the extravagance of jewelry, and perfectly shows out the brand's position in the market.

The door is designed with symmetrical way, and the symmetrical walls on both sides are hollowed and decorated with glass mirrors to form a display cabinet which can present some new products, so the store looks upscale and beautiful from the outside. The interior space is mainly decorated with wood to reflect the brand's concept of environmental protection.

NIE
DOVINIE

DOVINIE
BE DIFFERENT. BE DOVINIE.
DOVINIE
BE DIFFERENT. BE DOVINIE.

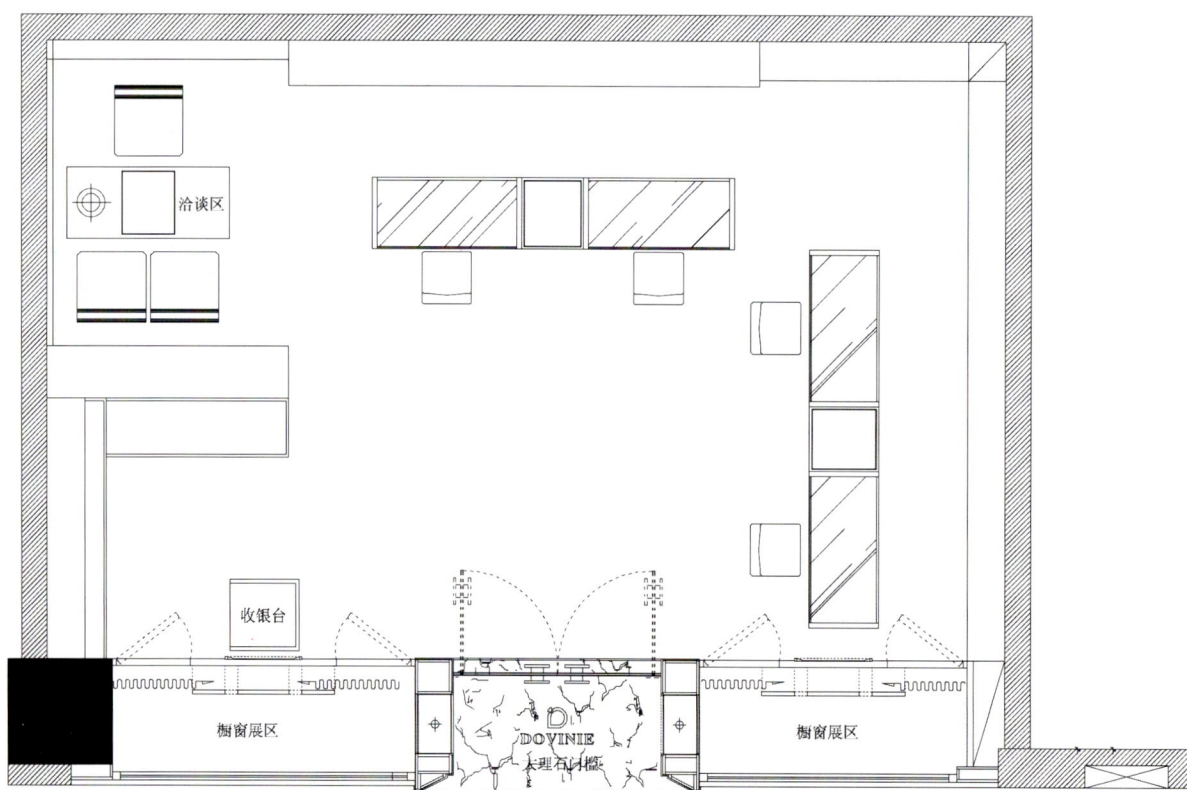

　　这是一个珠宝品牌店，室内的设计要求要有珠宝的贵气，还要能很好地体现出该品牌的市场地位。

　　门上采用两边对称式的设计手法，对称的两边墙面镂空并采用玻璃镜面装饰形成一个展示柜，里面可以展示一些新产品，让店铺的外表看上去高档、美观。室内多采用原木材料来体现品牌的环保理念。

Nine Silver Mill

九银坊

The brand is mainly engaged in sterling silver jewelry, so the store design reflects the high grade and pureness of silver jewelry.

The outside design introduces glass as the main material, combined with decorative veneers to design a push-and-pull door. The introduction of glass makes the space transparent, so people can feel the interior atmosphere from outside and could not help but want to come in to check it out. In addition, the design also makes good use of the load-bearing wall, where the brick paved wall is decorated to become the bright spot to reflect the interior grade.

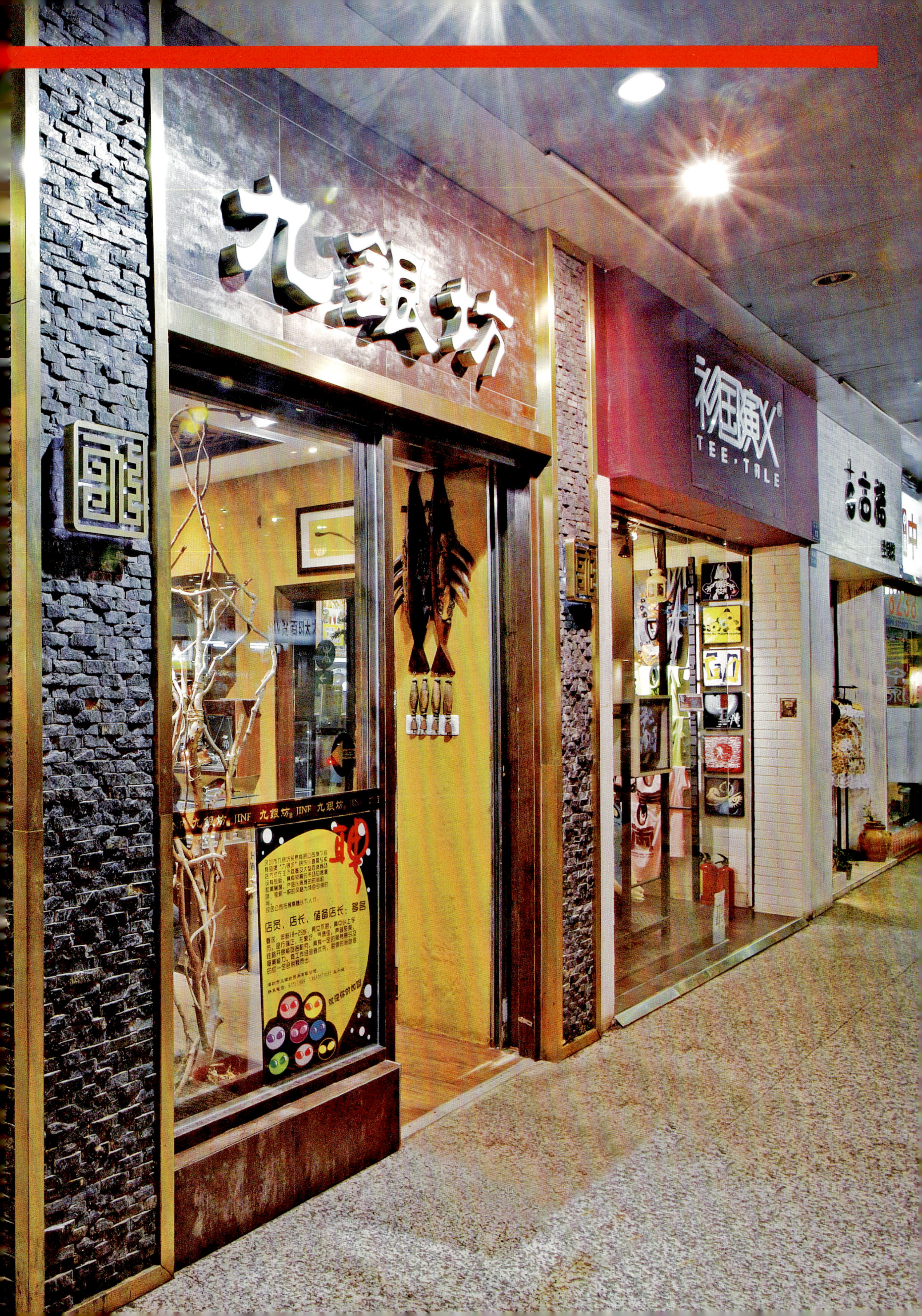
九银坊
衫国演义
TEE·TALE

该品牌主要经营纯银饰品，因此店内的设计要体现出银饰品的高档和纯净。

外观设计上，主要以玻璃为主材料，结合饰面板设计成推拉门。玻璃的运用让空间通透，让人在外面就能感受到店内的气氛，忍不住想进去一探究竟。此外，设计还很好地利用了承重墙，青砖铺贴的承重墙体经过装饰便成为了一个能够反映室内档次的亮点。

Elle's More Home Exhibition Hall   法视界家居展示厅

Cattelan Italia Home Exhibition Hall(Beijing)   意大利递展家居展示厅（北京）

Cattelan Italia Home Exhibition Hall(Shanghai)   意大利递展家居展示厅（上海）

Famous Mansion   名公馆

HC28 Home Exhibition Hall   HC28家居展示厅

LIVING

家居

地点	面积	设计师	软装设计师	设计公司	主要材料	摄影
/上海	/1000m²	/萧爱彬	/郭丽丽	/萧氏设计	/布艺	/萧爱华

Elle's More Home Exhibition Hall

法视界家居展示厅

"Choose freely and collocate with their own idea" is Elle's More's living concept for consumers, and the designers integrate the concept into the space design, allowing freedom to fill the whole space. The exhibition hall features two floors connected with each other by a winding stair, where the first floor is mainly for furniture, and the second floor mainly for accessories and outdoor furniture. The owner wants to operate with the way of humanization, so the designers add the display area for outdoor furniture on the second floor, which provides a place for display, communication and tea, integrating the display and interaction into a whole. The first floor is not divided by the hard decoration, and everything is done by the soft decoration to fully show the power of match. Furniture with different images and characteristics is rationally arranged together, independent and free to divide or combine. The cloth curtain at the entrance is designed with three colors, where gray, white and gold can be matched freely. The two white neo-classical droplights in the show window hanging down from the second floor, below it is the emmemobili flying saucer table, coupled with a fine selection of decorations and flowers, showing the elegance of a ballet dancer.

　　"选择自由，自主搭配"是法视界提供给消费者的生活观念，设计师也将这种观念融入空间设计中，让自由充斥整个空间。展厅一共两层，由一个旋梯连接，一楼以家具为主，二楼以饰品和户外家具为主。业主想以人性化的方式经营，因此设计师在二楼增加了户外家具展示区，既可展示又可交流、品茶，将展示和互动融为一体。一楼在硬装上几乎没做任何的分割，一切都从软装着手，充分发挥搭配的力量，将形象、特点各异的家具合理地组合在一起，既独立成区又分合自由。一进门的布帘采用三种颜色，灰、白、金可以自由调配，橱窗里两个白色新古典吊灯从二楼垂下，下面是emmemobili的飞碟桌搭配精心挑选的饰品和花，展示出了芭蕾舞者的优雅。

地 点	面 积	设计师	软装设计师	设计公司	主要材料	摄 影
/ 北 京	/ 216m²	/ 萧爱彬	/ 郭丽丽	/ 萧氏设计	/ 黑玻璃、发光灯片、宫廷灰大理石、橡木地板	/ 萧爱华

Cattelan Italia Home Exhibition Hall（Beijing）

—— 意大利递展家居展示厅（北京）

The generous and good manner of CATTELAN ITALIA makes the interior design of the exhibition hall must meet the neo-classical fashion and simple impressionistic style. The interior is comparatively concise, where the two black and white partitions are introduced to distinguish the neo-classical style from the modern one, and the deep-colored glass are used to divide and connect the two spaces to allow shoppers to find the object they want to buy quickly.

Noble and generous are the core quality of CATTELAN ITALIA, and the designers should strive to do so in interior design to match with the furniture. When you stand in the deep-colored space, the light-colored space in the distance attracts you to have a closer look; when you stand in the light-colored space, the furniture with metallic luster attracts you to come into another room, allowing you to wander in the store until you select the furniture satisfying you.

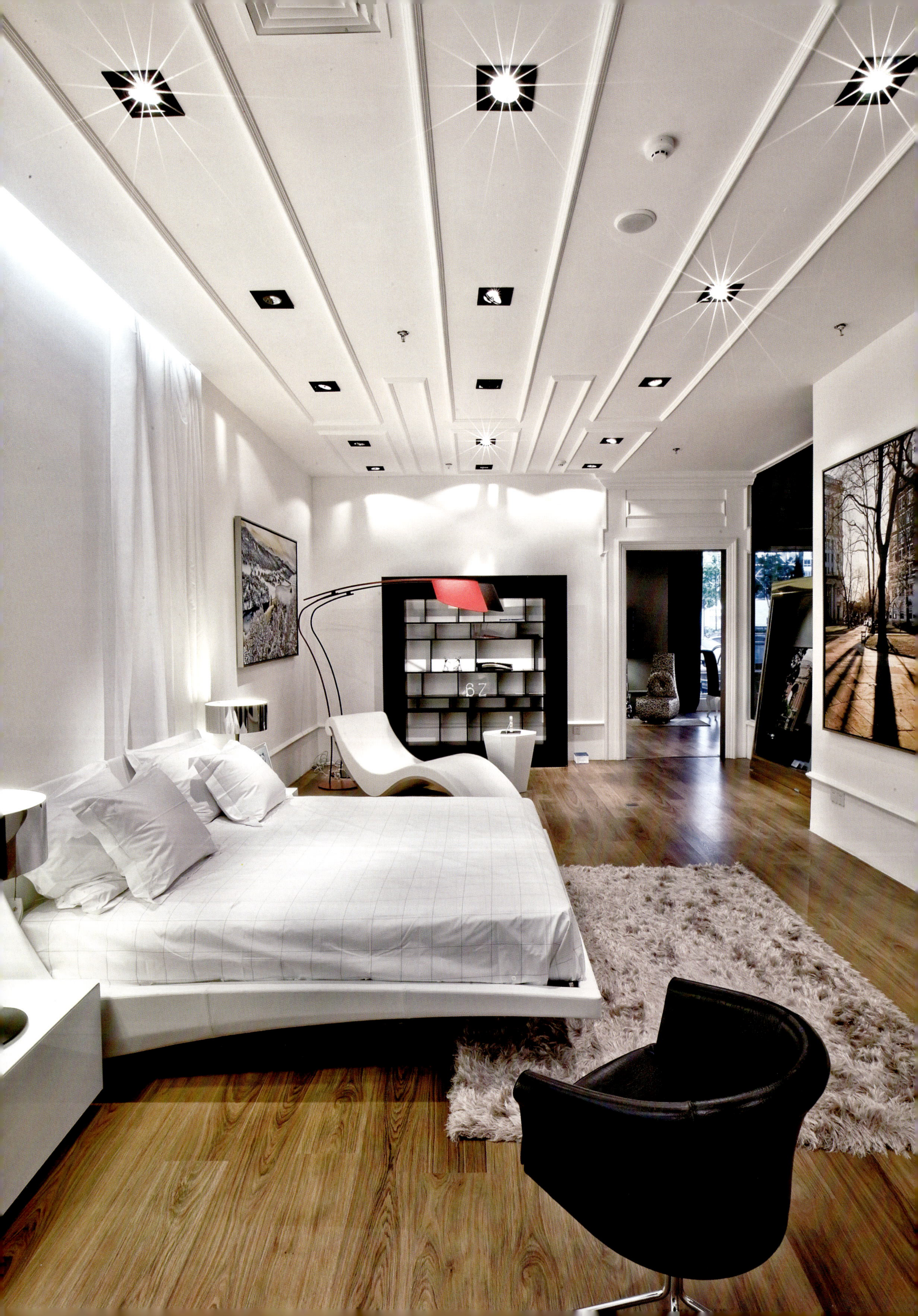

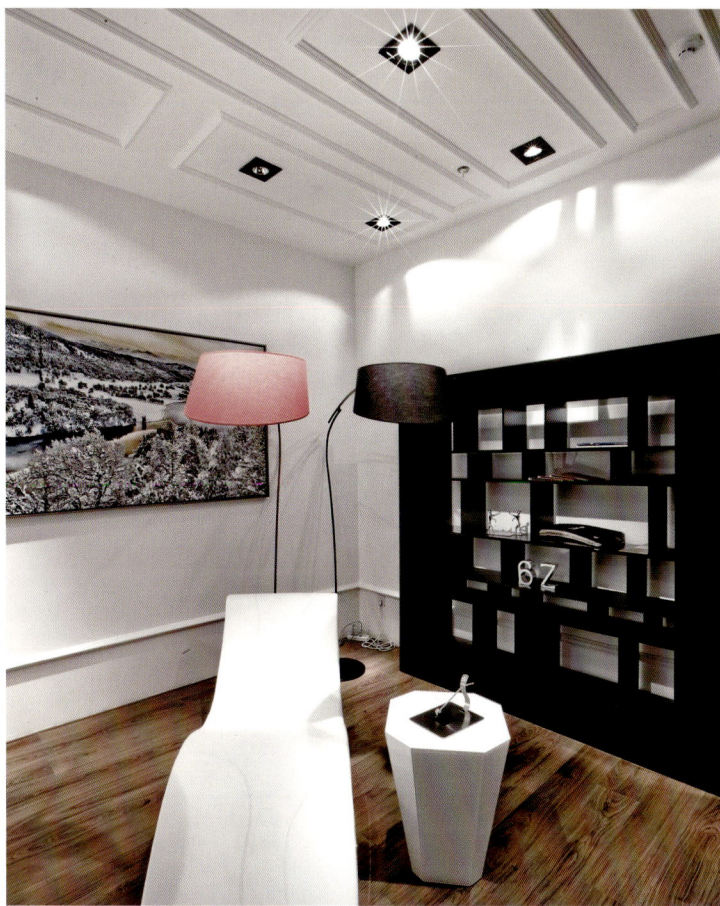

　　意大利递展家居大气而不俗的风度使得展厅的室内设计必须配合新古典主义的风尚和简约写意的风格。室内比较简洁，利用黑白两个分区把新古典主义和现代风格区别开来，深色玻璃又把两个空间既分开又联系，使购物者可以迅速到达自己想购买的商品处。

　　高贵而大气是意大利递展家居最核心的品质，室内设计要努力做到这点，才能与之相匹配。当您站在深色的空间时远处浅色的空间也同样吸引您去看个仔细；站在浅色空间里那带着金属光泽的家具又吸引着您进入另一个空间，让您在这店中徜徉，直到选到自己满意的家具。

地 点	面 积	设计师	软装设计师	设计公司	主要材料	摄 影
/ 上海虹桥	/ 280m²	/ 萧爱彬	/ 郭丽丽	/ 萧氏设计	/ 大理石、不锈钢、实木地板	/ 萧爱华

Cattelan Italia Home Exhibition Hall (Shanghai)

—— 意大利递展家居展示厅(上海)

The project mainly displays and sells Italian CATTELAN ITALIA and MAXDIVANI two brands of furniture, and "Cattelan Italia Home" is a new-style exhibition, consulting and sales platform for international home. It collects comprehensive first-line home brands with strong sense of design, refined workmanship, high quality and complementary product capabilities, perfectly fitting Chinese living demands and the value and aesthetics of high-end group, and remodeling a home space with exquisite design, elegant grade and superior cost performance.

The designers make contrast between neo-classical decoration and modern stylish furniture on purpose, to highlight the noble quality of Italian furniture. The ground and ceiling of the exhibition hall is designed with black and white in strong contrast, and a varied neo-classical door divides the space into two display areas. The floor lamps of unique style and the lighting shining on the stainless steel desktop form a reflecting effect, echoing up and down; the introduction of stainless steel further breaks the limitations of traditional Chinese materials of wood or glass. The exhibition hall is arranged according to the function of furniture, so the themes of each section are clear, not only artistic, but also reflects the practical function.

本案主要展示和出售意大利CATTELAN ITALIA和MAXDIVANI两个品牌的家具。"递展家居"是新型的国际家居展示、咨询和销售的平台。在国际上搜寻多个设计感强、做工精致、品质上乘、产品功能互补而全面的一线家居品牌，完美贴合中国的居住需求以及高端人群的价值观和审美观，重塑了一个设计精妙、品位高雅、性价比卓越的家居空间。

设计师有意用新古典主义装饰和现代的时尚家具形成对比，借此强调意大利家具高贵的品质。展厅的地面和顶面用了对比鲜明的黑白两色，一个变异的新古典主义的大门，分隔了前后两个展示区。造型独特的落地灯，灯光照耀在不锈钢的桌面上，形成了反射效果，上下呼应；不锈钢材料的应用更是打破了中国传统木制或者玻璃材料的局限。展厅是按照家具的功能来布置的，因此每部分的主题都一目了然，不仅有艺术性还体现了实用性。

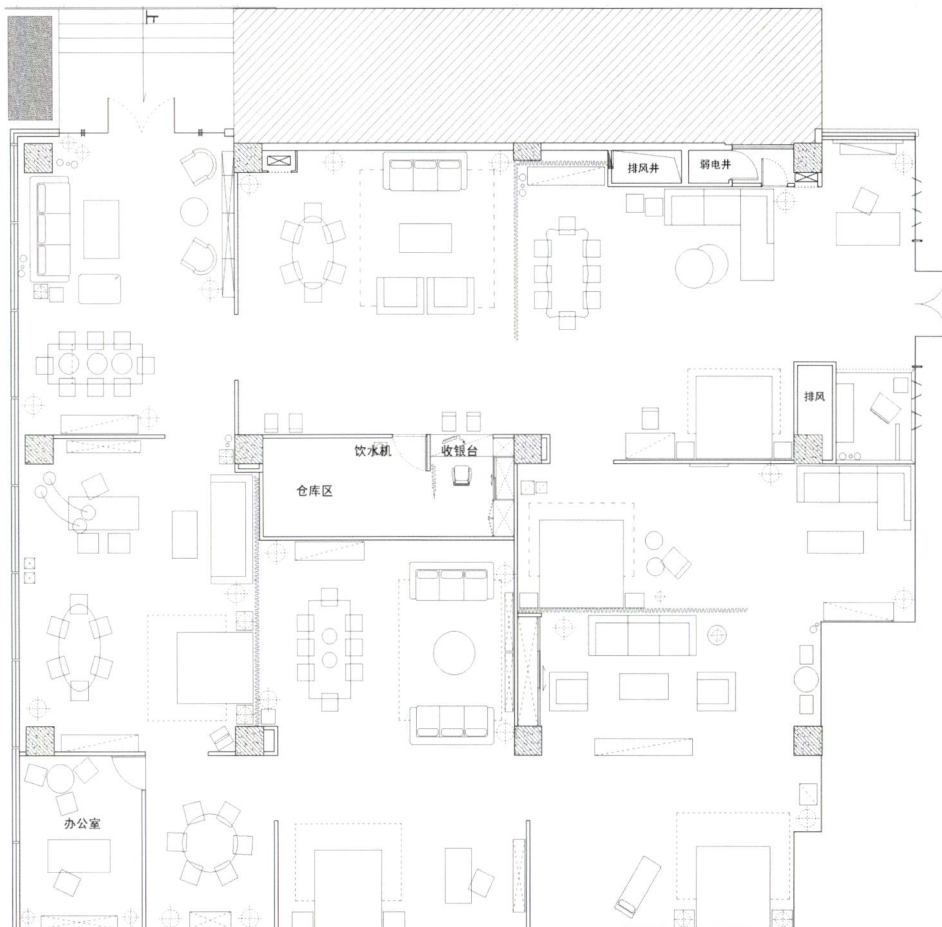

地点	面积	设计师	设计公司
/ 汕头	/ 1150m²	/ 李伟光	/ 汕头市丽景装饰设计有限公司

Famous Mansion

___ 名公馆

In the project of Famous Mansion, the designer introduces the soft-decorated whole matching concept throughout the whole design. How to allow customers to understand the design, furniture, wallpaper, curtain and other soft decorative materials most easily and directly is the focus of the design, meanwhile, he customizes hard-decorated environment for each set of the furniture. The materials of wallpaper and curtain, color collocation and the exquisite detail of light source allow the displaying effect of the whole physical scene to make customers feel like in their new home, and this feeling of being personally on the scene makes customers assured and satisfied.

Famous Residence is positioned to be a high-end concept store of soft-decoration one-stop matching, not only making it convenient for the designer to design, but also providing an exhibition hall of comprehensive, multi-style and multi-model show flat for customers. The real scene display allows customers to feel the nature of the furniture and the space environment, featuring a close contact with the color and texture of materials.

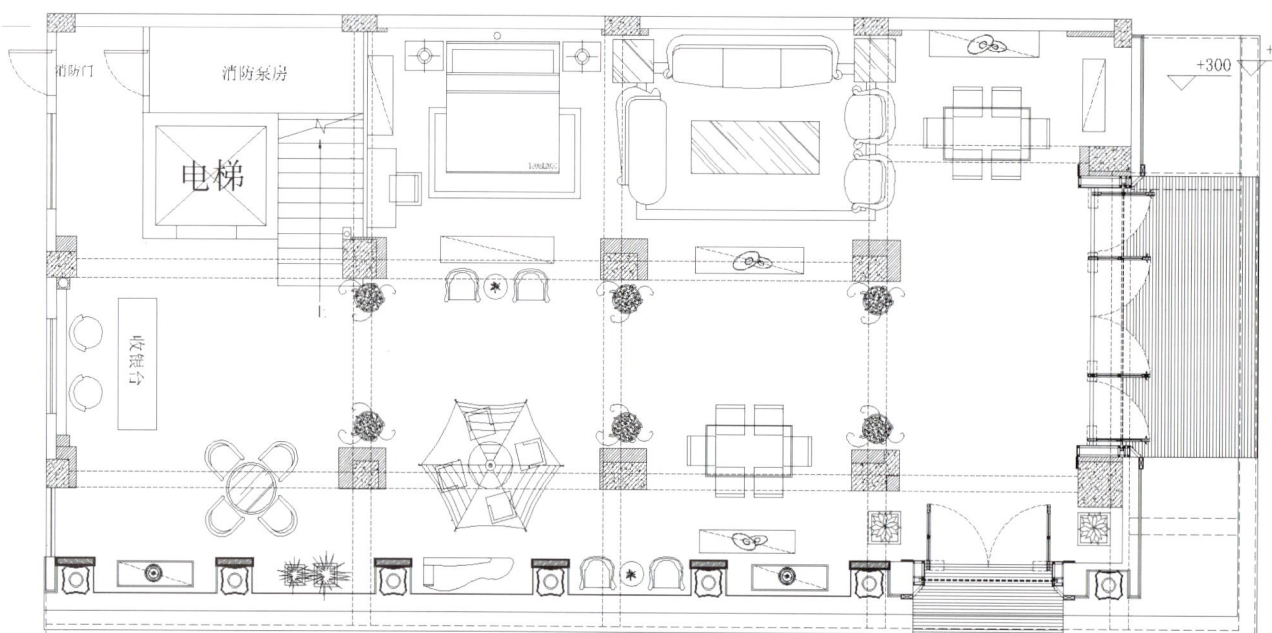

　　名公馆项目中，设计师将家具软装整体配套概念贯穿于整个设计中，将如何使业主最容易、最直观了解设计，了解家具，了解墙纸、窗帘等软装素材作为设计的重点，为每套家具度身定制硬装环境。墙纸、窗帘的材质，色调的搭配及照射光源的细致考究，使整个实物场景的展示效果让业主完全可以感受到自己的新家就是这样一种感觉和场景，这种身临其境的感觉可以让业主百分之百的放心和满意。

　　名公馆定位高端家具软装一站式配套概念店，这不单为设计师的设计工作提供了方便，更为业主提供了一个全方位、多风格、多款式的样板间大汇展。真实的场景展示能让业主感受到家具的内涵和空间的氛围，还能亲密接触材料的色调和质感。

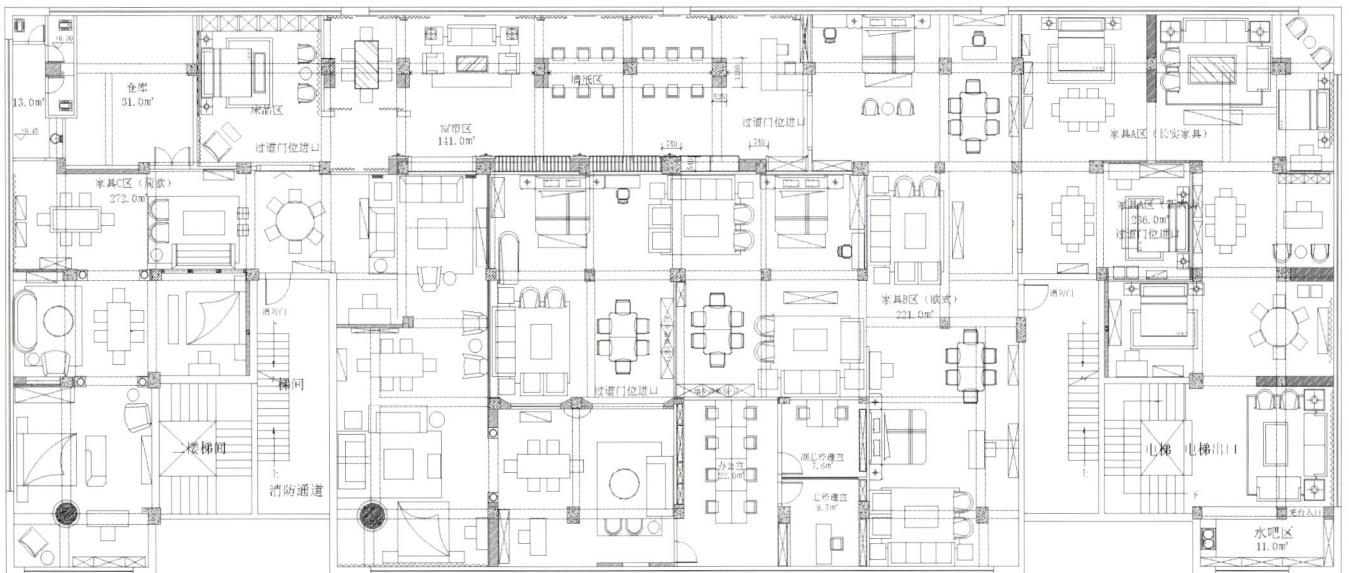

地点	面积	设计师	软装设计师	设计公司	主要材料	摄影
/ 宁波	/ 210m²	/ 萧爱彬	/ 郭丽丽	/ 萧氏设计	/ 有机玻璃、订制白色复合地板	/ 萧爱华

HC28 Home Exhibition Hall

___ HC28家居展示厅

Once the imported furniture becomes the fashion, various brands have opened stores across the whole nation, and "HC28" is an example. The designers give the design a style position which is different from the ordinary sense of "modern". In HC28 home, the "simple and rustic space" is coupled with the "oriental modern style furniture" to allow "simple and pure" and "gorgeous and rich" to reflect and shine mutually, creating a relaxing and peaceful Oriental romance.

The space division of the whole furniture store is in modern free form, where the division of "dynamic" and "static" makes each area relatively independent without disturbing each other. The whole color of the space echoes with the colors of furniture, and the deep gray, white, black, wood color and other colors are used together to show the integrity of the space.

"The function and beauty should be integrated, and it's hard to say the pros and cons, we should take both of them into account." Space can produce dialogue with human mind, and a good interior space can touch us. The design of the furniture store carries the purpose of stimulating consumption; therefore, the designers are required to think in depth to touch people's hearts.

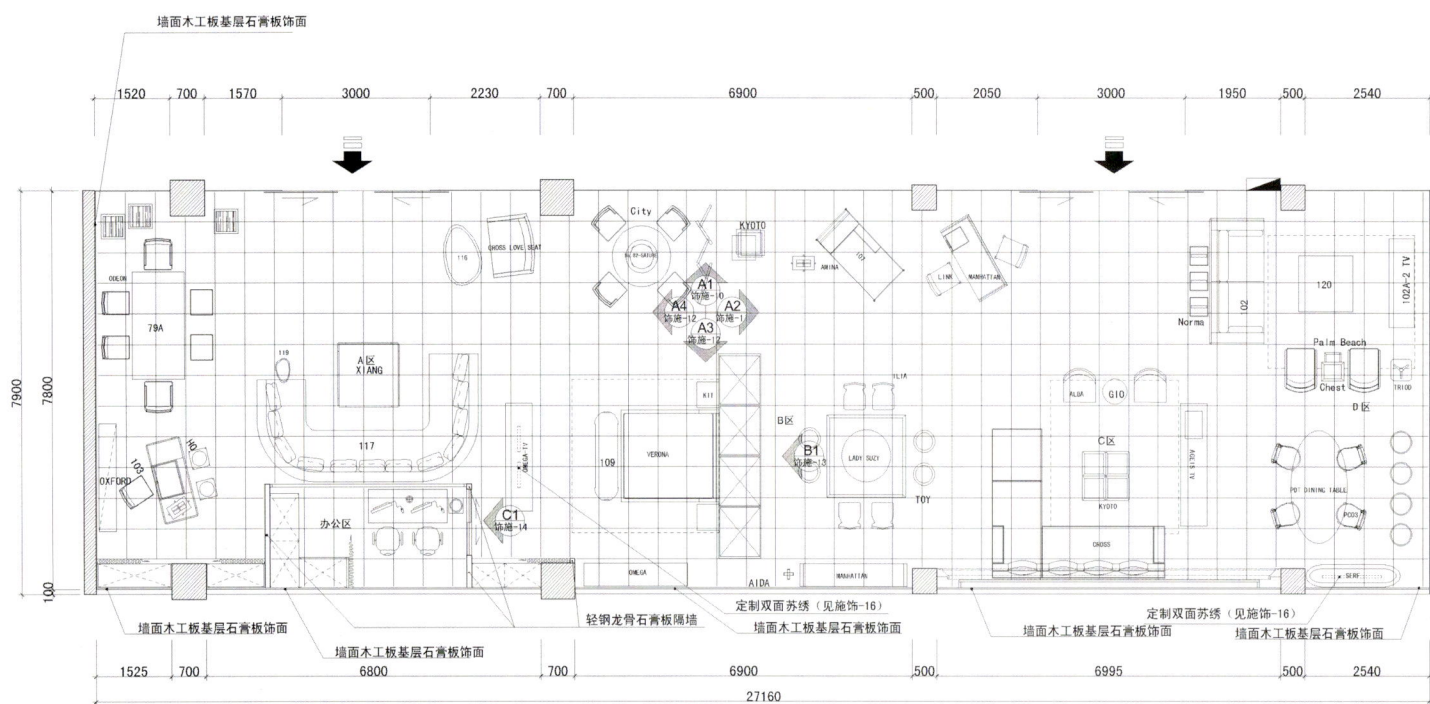

墙面木工板基层石膏板饰面

1520　700　1570　　3000　　2230　700　　　　6900　　　　500　2050　　3000　　1950　500　2540

7900
7800
100

ODEON
79A
OXFORD
办公区
C1 饰施-14
A 厢 XIANG
117
119
118
CROSS LOVE SEAT
City
KYOTO
A4 饰施-12　A1 饰施-10
A2 饰施-11
A3 饰施-12
OMEGA TV
109
VERONA
K11
B 厢
B1 饰施-13
OMEGA
AIDA
MANHATTAN
LADY SUCY
TOY
FL1A
ALDA
G10
AG13 TV
C 厢
KYOTO
CROSS
LINK
MANHATTAN
Norma
102
Palm Beach
Chest
D 厢
TR100
POT DINING TABLE
PC03
SERF
120
102X-2 TV

墙面木工板基层石膏板饰面
墙面木工板基层石膏板饰面
轻钢龙骨石膏板隔墙
墙面木工板基层石膏板饰面
定制双面苏绣（见施饰-16）
墙面木工板基层石膏板饰面
墙面木工板基层石膏板饰面
定制双面苏绣（见施饰-16）

1525　700　　　　6800　　　　700　　　6900　　　500　　　6995　　　500　　2540

27160

　　进口家具的旋风一经刮起，很多品牌都在全国各地开出分店，"HC28"就是一例。设计师对于此设计的风格定位区别于普通意义上的"现代"。"简约质朴的空间"搭配"东方的现代风格"的HC28家具，让"简洁纯粹"和"华丽浓郁"交相辉映，打造出放松与平和的东方浪漫。

　　整个家具店的空间划分采取现代自由的形式，"动"与"静"的划分使各个区域互不干扰，相对独立。空间整体的色彩运用和家具的用色相呼应，运用深灰色、白色、黑色、木色等色调共同体现空间的整体性。

　　"功能和美观应该融为一体，很难说孰轻孰重，要兼顾考虑。"空间可以和人的心灵对话，好的室内空间能打动人心。家具店的设计承载着刺激消费的目的，更要求设计者进行深入思考，触动人们的心灵。

Rebecca 瑞贝卡

Bees Ceramic Tile Concept Store ″蜜蜂瓷砖″概念店

Apodon In Xiamen 厦门光合作用

JNJ Mosaic Store In Jockey Club, Guangzhou JNJ马赛克广州马会专卖店

OTHERS

其他

地 点	面 积	设计师	参与设计	设计公司	主要材料		摄 影
/ 北京	/ 200m²	/ 利旭恒	/ 王 鹏	/ 古鲁奇公司	/ 金属帘、镜面、陶瓷片、玻璃纤维、白色地砖		/ 孙翔宇

Rebecca

___ 瑞贝卡

Rebecca is the largest wig manufacturer in the world, and it's also a world-class brand. The designers choose the structure and curve of women's hair as the designing concept of the store, providing the customer a shopping environment which is full of affinity. It makes wig a fashion accessory, just like hats or boots, giving customers a fresh opportunity to understand and contact with wig products.

Rebecca是全世界最大的假发生产商，同时也是国际级的品牌。设计师将女人发丝的结构与曲线作为专卖店的设计概念，提供给大众一个亲和力十足的购物环境，让假发成为像帽子或靴子一样的时尚配件，给人们一个全新的机会来认识与接触假发产品。

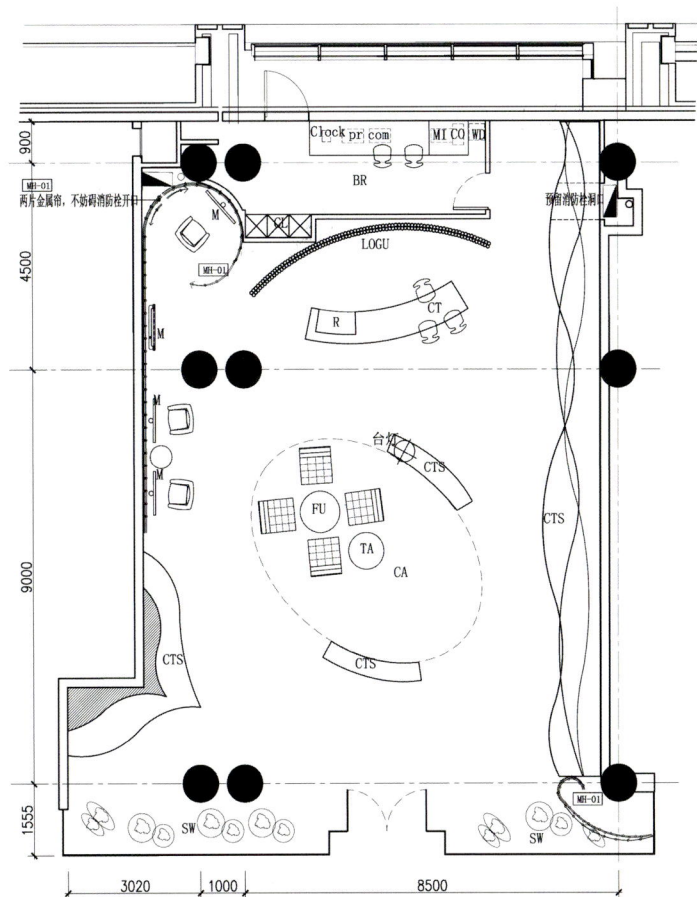

com	电脑
pr	打印机
Clock	打卡机
SA	小型保险柜
CO	咖啡机
MI	微波炉
WD	饮水机
CL	更衣柜
LOGU	主题背景墙
BR	工作后区
R	收银台
CTS	展示
CT	柜台
TA	花卉展台（外购）
M	试戴区镜面
DPT	展示桌（外购）
MH-01	金属帘隔断
CA	定制地毯
SW	橱窗展示
FU	成品家具（外购）

Bees Ceramic Tile Concept Store

—— "蜜蜂瓷砖" 概念店

Proposition Orientation

As the most important brand of Italian Imola, "Bee" Ceramic Tile features more than 100 years of history from starting, and it has been famous in Europe for its innovative design. After nearly 10 years of development in China, it has become a high-profile brand of imported building materials.

The concept store is designed to enhance the brand's displaying image in Chinese market, and strengthen the product features, combined with China's national conditions and consumption characteristics to integrate "international, fashionable, classic, qualified, professional" and other elements into the design of exhibition space.

Environmental Style

The brand features the biggest selling point of rich colors and varied patterns and textures. In order to express the nature of goods better, the designers put the light environment design on a strategic location. From the lamps selection, light source determination to the using of the latest computer technology for simulating test, to the late adjustment, the designers do the repeated and meticulous researches in order to achieve a dynamic, flexible and controllable lighting environment. Light is introduced as a tool to emphasize the art, dramatic and focused commodity, creating a comfortable shopping space and building a more accurate and reliable communication relationship between goods and men.

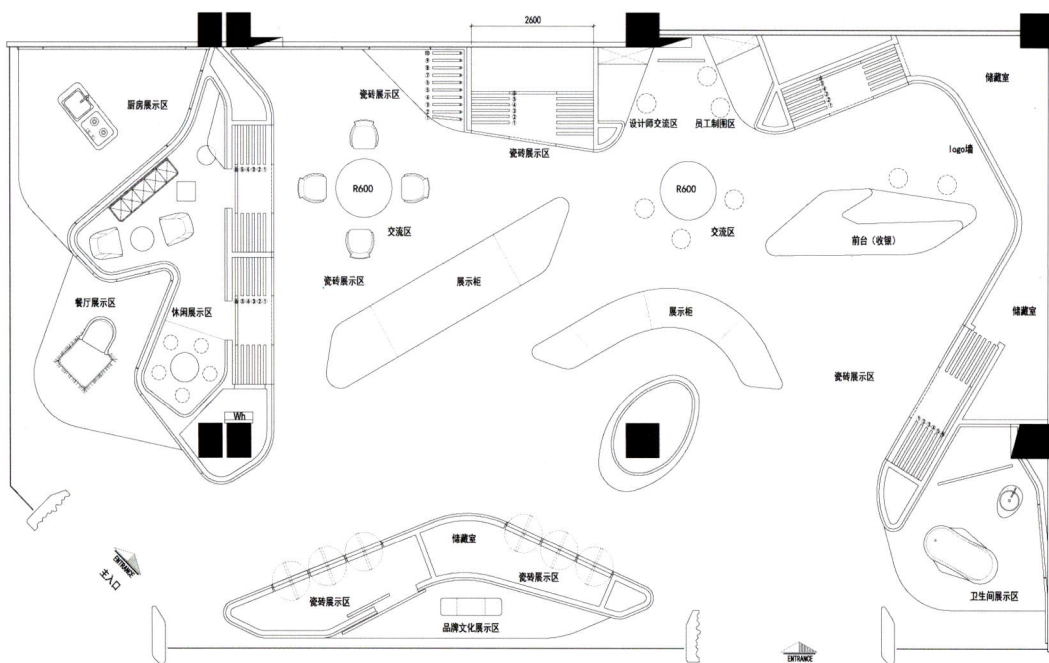

项目定位

"蜜蜂"瓷砖作为意大利伊莫拉公司旗下最重要的品牌,从创业至今,有着100多年的历史,一直以创新设计著称于欧洲,在中国经过近十年的发展已使其成为知名度极高的进口建材品牌。

此次概念店的设计目的是提升品牌在中国市场的展示形象,强化产品特征,结合中国国情及消费特点,将"国际、时尚、经典、品质、专业"等元素导入到展示空间设计中。

环境风格

该品牌商品以丰富多彩的颜色和多变的花样、肌理为最大的卖点。为更好表达商品的特质,设计师将光环境的设计放在了战略性的位置。从灯具选型、光源的确定到利用最新的电脑技术进行模拟测试,到后期的调整等都进行了反复的细致入微的研究,以营造动感、灵活、可控的照明环境。以光为工具,强调商品的艺术化、戏剧化、焦点化,打造舒适的购物空间,在商品与人之间建立更准确、更信赖的信息沟通关系。

IMOLA

地点	面积	设计师	设计公司	摄 影
/ 厦 门	/ 950m²	/ 迫庆一郎、原信敏、长门宏明、渡边修一	/ SAKO建筑设计工社	/ 许晓东摄影工作室

Apodon In Xiamen

—— 厦门光合作用

Bookstore space consisting of organic "landforms"

The design concept is inspired by the name of the bookstore "APODON". The "sky shelves" and "ocean shelves" showcases hanging from the ceiling, "grain shelves" on the wall floating with wind and "tree shelves" around pillars, these organic "landforms" is dotted with green scindapsus to fill the whole space with a sense of life. Between the bookstore and customers, the "photosynthesis" between knowledge and oxygen is quietly playing a role.

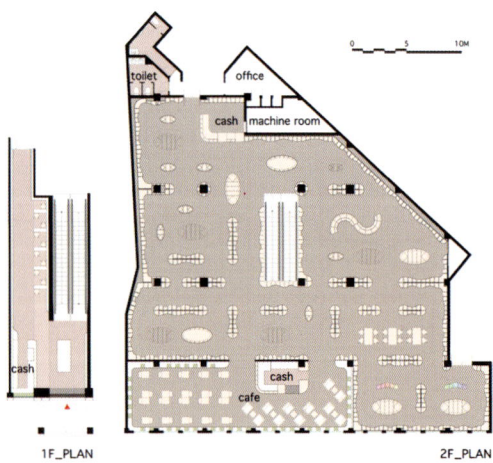

由有机"地貌"组成的书店空间

　　设计的理念来源于店名"光合作用"。悬挂在天花板上的"天空书架"、"海洋书架"展示台，墙壁上随风浮动的"谷物书架"以及柱子周围的"树木书架"，这些有机的"地貌"加上绿萝的点缀，使整个空间充满生命感。书店和顾客之间，知识与氧气的"光合作用"正在悄然发挥作用。

地 点	设计师	参与设计	设计公司
/ 广 州	/ 谢智明	/ 霍律鸣、叶锦波	/ 大木明威社建筑设计有限公司

JNJ Mosaic Store In Jockey Club, Guangzhou

JNJ马赛克广州马会专卖店

JNJ mosaic Guangzhou Image and Concept store is located in the eastern zone of Jockey Club Home, Pearl River New City, Tianhe District, Guangzhou. The space limitations and the narrow storefront make it more difficult to design, and in the project, the designers cleverly introduce the space concept of egg-roll shaped space-time tunnel to design the space, breaking the limitations of the long and narrow space from inside to outside; at the same time, with the advantages and artistic quality of mosaic inlaid materials, the designers show the space concept and practicality of the whole mosaic image store. The whole store is divided into two areas, concept space display area in the front and select and service area at the back, with the reasonable space division and the mix-and-match way of expression to show a mosaic brand store with the integration of artistic quality and function.

234

　　JNJ马赛克广州形象概念店，选址在广州市天河区珠江新城马会家居东区。空间的局限性及狭长的店面为设计增加了难度，本案中，设计师巧妙地运用蛋卷形时空隧道的空间概念来设计空间，由内而外地打破了狭长的空间带来的局限性；同时还结合马赛克镶嵌材质的优势及艺术性来表现整个马赛克形象店的空间概念及实用性。整个店面分前区的概念空间展示区及后区的选样服务区，以合理的空间分隔及混搭的表现手法，展现了一个艺术性与功能性相结合的马赛克品牌专卖店。

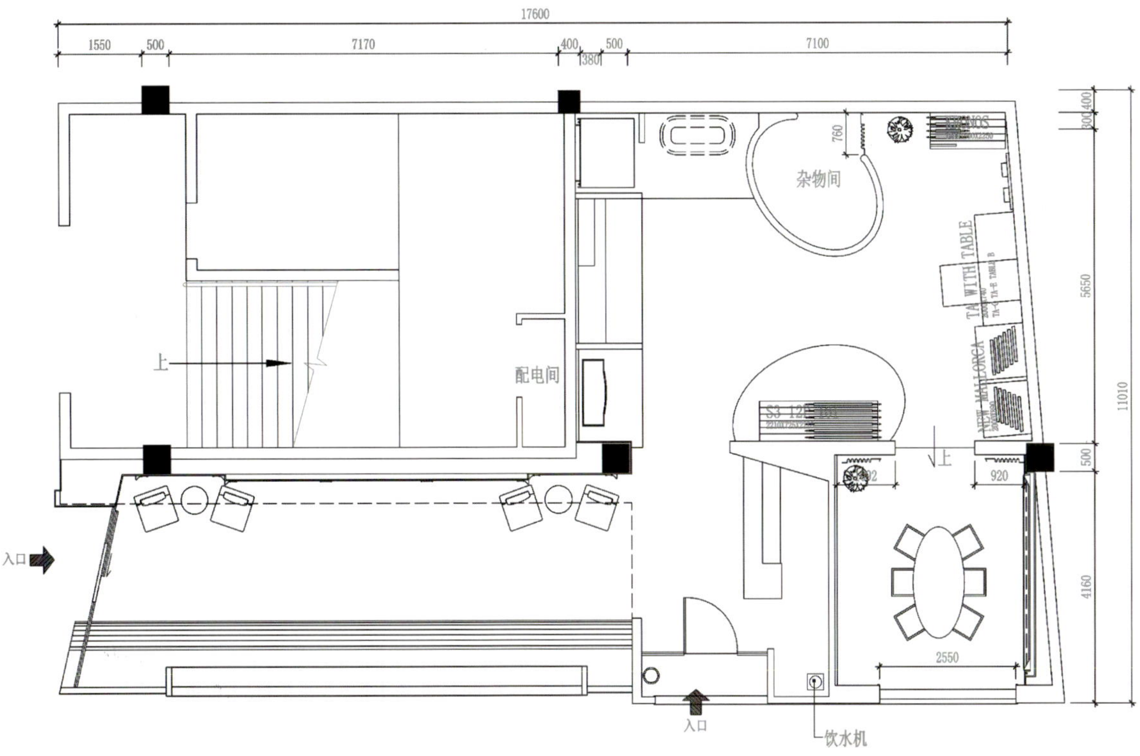
17600
1550
500
7170
400
500
7100
380
760
杂物间
配电间
上
入口
5650
11010
500
920
4160
2550
饮水机
300
400

本书参编人员

周锋、卢霭潮、刘宝达、欧阳亮、周强、陈哲、周美龄、雷小兰、胡青、吴俊、方丽、段君龙、周晓琪、庄丽娟、周琴、赵丹、赵标、闫兴宝、徐剑、王琪、黄芸、孙峰、黄宗坤、王雪松、贾春萍、李红靖、黄静、黄康裕、杜小慧、吴俭英

参与本书翻译的人员

范连颖、王丽红